Fertigation Scheduling of Field Crops

NIPA GENX ELECTRONIC RESOURCES & SOLUTIONS P. LTD.
New Delhi-110 034

Similar Book of Interest

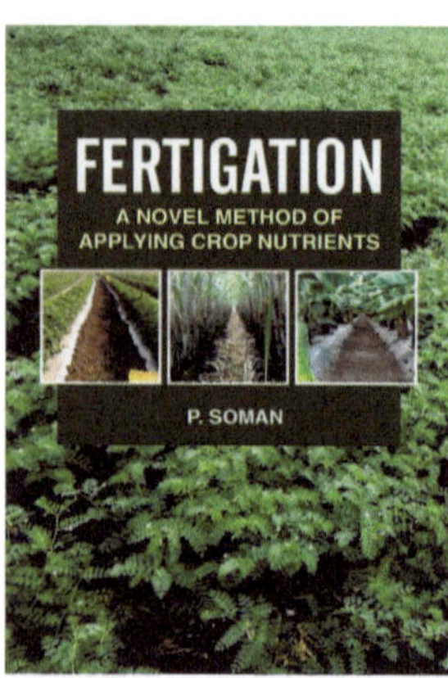

Fertigation
A Novel Method of
Applying Crop Nutrient

P. Soman

ISBN: 9789390591145

Year: 2021

Fertigation requires a thorough understanding of the science behind the technology to make it deliver the immense possibility it offers in crop production. Though the idea of fertigation existed from the times of solution culture, it did not receive the necessary attention from among plant nutritionists and agronomists when it reappeared in the context of micro irrigation.

Fertilizer application in field agriculture has also not developed as a precision technology. Recommendations of the quantum of fertilizers required for a crop, at least in India are not based on current varieties of the crops, nor have they anything to do with the growth rate and developmental changes occurring while a crop is managed by the grower. Most of the fertilizer recommendations are itself very old and efforts to make them relevant to the current growing conditions, soil status, crop variety and crops reaction to the environment etc. are very limited. It is even worse when growers follow traders' recommendations whose idea is to sell more the fertilizer they supply. Not only lower yields and very low fertilizer use efficiencies, but the deterioration of soil and water bodies are the results.

Fertigation Scheduling of Field Crops

P. Soman
Chief Agronomist
Jain Irrigation (Global)
Jalgaon, Maharashtra, India

NIPA GENX ELECTRONIC RESOURCES & SOLUTIONS P. LTD.
New Delhi-110 034

NIPA GENX ELECTRONIC RESOURCES & SOLUTIONS P. LTD.

101,103, Vikas Surya Plaza, CU Block
L.S.C.Market, Pitam Pura, New Delhi-110 034
Ph : +91 11 27341616, 27341717, 27341718
E-mail:newindiapublishingagency@gmail.com
www: www.nipabooks.com

For customer assistance, please contact
Phone: + 91-11-27 34 17 17 Fax: + 91-11- 27 34 16 16
E-Mail: feedbacks@nipabooks.com

ISBN: 978-93-91383-34-3

Composed and Designed by NIPA.

Preface

Scheduling means *Which nutrient? How much?* and *When* to apply?

As part of providing technical support to micro irrigation adopted farmers, I have spent time to understand more about fertigation and been preparing schedules to give to the adoptees for many crops-cereals, pulses, oil seeds, commercial non-food crops, seasonal vegetables, field flowers and fruits and perennial fruits and other crops. Several training programs are conducted for Agronomists and Farmers in various countries where micro irrigation systems are established by engineers. Because the very concept of fertigation is under-studied and ill-researched in India, it is all the more important to publish some of the basics of the process and provide a ready to use guide. With this thought process, first a book on the basics was published in 2020. While this second book on the subject gives a guide on scientific scheduling of fertigation for as many crops as I have worked so far.

The basic quantum of NPK used to prepare the schedules are actually the recommended doses (RD) of NPK by relevant public institutions or state agriculture universities. Using these basic recommendations, schedules are prepared to match the growth and the known developmental stages of each crop. In order to enhance the use of the schedules to other regions/countries where different RD is in practice, I have worked out the basic nutrient split ratios for each growth stage duration. Using these ratios the fertilizers are to be partitioned for each stage of the crop. Each of these parts of fertilizers are then converted into farmer-usable daily doses. However, the schedule is not given on a daily application basis as it was found that most farmers do not want to go for daily dosing (injection). Hence the schedule is reported using a dose for every three days. It is also found that more frequent application schedule will give better yield and nutrient use efficiency.

As we introduce drip and fertigation to further more crops schedules are also prepared for the new crops. The schedules of most of the crops given in this book are actually field applied by farmers during the last two decades and been in use for several seasons.

Author gratefully acknowledges Jain Irrigation Systems Ltd. for providing an opportunity to work with farmers that actually led to the circumstances to study about this new technology of fertigation and apply real-time solutions.

Author

Contents

1

Introduction

While fertilizer injectors that mixes fertilizer solutions with the irrigation water flow in the micro irrigation systems, a scientifically prepared fertigation schedule is the software that farmers need to implement a successful fertigation plan.

Though water soluble fertilizers are a special grade of fertilizers manufactured for fertigation *per se*; I have been assisting farmers to do fertigation with conventional fertilizers, those are familiar to them; like Urea for N, white muriate of potash (MOP) for K. Only in the case of P source they need to acquire water soluble special fertilizers like Monoammonium Phosphate (MAP) or Phosphoric acid. Historically, fertigation by our small holder farmer began in India just by fertigating Urea alone. And applying P and K as direct dry application to the soil with Single Super Phosphate (SSP) or pink MOP, both are not water soluble. Gradually, when farmers realised that fertigation is giving them more yield of the crop and income they warmed up to the idea of full fertigation and demand for water soluble fertilizers like MAP, White Potash, SOP etc. grew.

In Indian situation some 20 years back when I started advising farmers on fertigation as part of drip irrigated precision farming, I first took only Urea for fertigation. Later MOP (white) was added as a K source. This continued for a number of years with P applied as Basal dose to the soil directly. Later progressive farmers began asking for fertigation plan for P also for many highly profitable horticulture crops. Then the whole fertilizer application fell into shape of fertigation.

The other major factor that would impact on the adoption of fertigation is the cost of water soluble sources of N, P or K. This has led to making recommendations of fertigation schedules which included conventional fertilizers like Urea and water soluble fertilizers like MAP, Phosphoric acid, white MOP or balanced water soluble complexes like 19:19:19 etc. This strategy has worked well. More and more farmers found this is best for them from the point of view of cost and getting high yields. Therefore, the schedules of different crops given in this guide book follows that strategy. It proved to bea very practical and farmer friendly strategy.

Factors considered while preparing a scientific schedule

Success of drip fertigation technology depends mainly on the plan (schedule) of irrigation and fertigation-those that the farmers are advised to follow. The value of a drip system and its impact on crop production is solely due to correct schedules prepared scientifically. In many cases, as I have seen in Indian small farmer communities, many of them are not provided with a fertigation schedule at all after installing a drip system. There is a lot of confusion and lack of understanding when it comes to scheduling of fertigation. Even crop specialists tend to prepare fertigation doses simply by dividing the total fertilizer quantity by the number of days of crop growth. This is highly erroneous.

Frequency of fertigation

How often the farmer should do fertigation. If left to him he may apply only 4 or 8 times during the crop duration. Here the technology adviser/extension person have to intervene and educate the farmers about more frequent application. I have been asking them to fertigate daily or once in 2 or 3 days. This is the best to get maximum fertilizer use-efficiency, yield and income. In case of seasonal crops, Rice, Wheat, Cabbage, Potato etc. this is very important that they are fertigated at least once in 3 days. A certain level of laxity can be accommodated in case of perennials where changes of growth stages are relatively less rapid. Hence I have taken the frequency as once in 4 days (2 times a week). The schedules in this guide are made in such a way that even if a farmer prefers to go for a lower frequency he can do that by adding 2, two day doses into one. However, frequencies more than a week per dose of fertilizer is not found to be useful and productive.

Fertilizer recommendation

As much as possible published recommendations for crops by state universities or government institutes are followed to get the total quantum of NPK for each crop. Any farmer/agronomist wanting to use the schedule in another location with different NPK recommendation can easily do that by following the ratio of nutrients for each growth stage (duration) of the crop. Use the ratio and prepare schedule using the local recommendation for that crop.

Nutrient requirement plan at any one time in a crop cycle depends on crop characteristics

1. Expected final yield/ yield target.
2. Nutrient content in the harvested part and the residual biomass
3. Environmental conditions- temperature, humidity and light.
4. Soil nutrient status (for soil grown crops)

Considering all these factors specific nutrient recommendations for a crop have to be based on nutrient uptake measurements under conditions as near as possible to those in which the crop is grown. However, this is not always possible especially in the innumerable villages located in large districts of several states of the country. Often times the nearest location from where a recommendation is made can be 400-500 km away from a small farm where the crop is grown.

Therefore it is more or less accepted that generalized fertilizer recommendations can be utilized for fertilization plan for a specific crop and its different cultivars. Nonetheless, fertigation gives a tool to make regional modifications of the fertilizer quanta and timing of application based on soil and leaf/petiole analysis. A far rapid method of adjustments in the application rate achieving precision is possible.

Co-ordination of supply of nutrients and changing nutrient demand of the crop

A scientifically prepared fertigation plan allows coordination of nutrient supply with changing demands of the growing crops which changes with the physiological shifts in the crop (growth phases). Fertigation planner requires following information to prepare the schedule- *which, when and how much* plan:

1. Crop growing cycle with development stages and their duration
2. How the crop growth cycle is influenced by local climate.
3. Amount of each nutrient required by the crop at each stage
4. Nutrient removal pattern of the crop from the soil/media (Fig. 1)

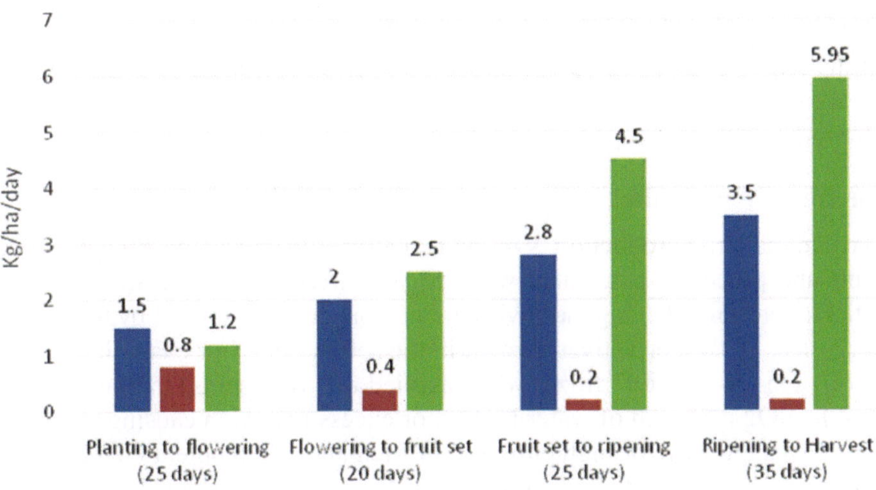

Fig. 1. Nutrient uptake rate recommended based on the uptake rate of Field Tomato – *Data Source: International Potash Institute (IPI).*

1. Crop Growth Stages

A crop goes through many stages during its growth from germination to fruit production and harvest. Nutrients are absorbed, transported and utilized during these stages till the crop arrives at physiological maturity beyond which there is no growth (seasonal and annual crops); or fruit formation and harvest followed by repeat vegetative stage (perennial crops). To prepare a fertigation schedule one should have a clear idea of these stages and their durations. Nutrient requirements vary both in type and quantity during these growth stages.

Example - Banana

i) Early vegetative growth

ii) Late vegetative growth

iii) Bunch exit

iv) Bunch and finger growth

Fertigation schedule normally prepared from the local fertilizer recommendations should be adjusted from time to time by:

1. Making regular observation of the crop
2. Identifying any symptoms of nutrient deficiency
3. Analyzing the leaf/petiole and recalibrating the fertilizer quantity and type in the schedule.
4. Adjusting the schedule for any season elongation or change in the developmental phase of the crop.

Availability of nutrient ratios with respect to the physiological stage of the crop would help in arriving at the correct fertigation schedule.

Throughout this book, growth stages are not named but their durations are given to make it more farmer friendly. It will help farmers to understand the schedules in terms growth time (day) from planting.

Benefits of Fertigation

Conventional crop production is found to add to climate change phenomenon. Inputs and processes associated with irrigation, fertilizer application, use of pesticides etc. are affecting the environment more detrimentally; 1 by the inputs *per se* or 2. by the derivatives formed in the soil system; For example, excess use of Urea (N-source fertilizer when applied in 1 or 2 splits) in rice paddies result in NO2 pollution of water bodies; or excess irrigation causing standing water paves the way for high emission of Methane gas, a potent GHG. When fertigation is practised the quantum of fertilizers reaching the soil at any one time is very little and will be in plant absorbable form thus avoiding the GHG formation.

Fertigation helps in actual reduction of fertilizer load on to the soil still increasing crop's yield.

- Fertigation with a scientifically designed schedule allows nutrient application in the correct dosage at appropriate time of crop's growth.
- Very low rates of nutrients at the early growth stage
- Rates increases as the flower and fruit load increases
- Again rates reduce as the crop goes through senescence and approaches end of growth.
- Crop gets its requirement satisfied all through its growing cycle and hence yields optimally.
- The nutrient use efficiency of the crop is improved by fertigation.

Because of the above strict regimen soil is never loaded with excess fertilizer. This alone is adding to lower rate of residual accumulation in the soil and ground water.

Challenges in the adoption of fertigation by growers

1. Low farmer awareness of the process and its benefits

Solutions: More focussed extension program and farmer training at field level is a requisite. This should be done through departments and officers in the department should be trained first. There should be at least two fertigation trained personnel for each block. Some private supplier companies can provide training through their experts.

Just like when the use of chemical fertilizer was introduced in the country, we have to take up promotion exercises of fertigation among the small farmers. Introducing fertigation science and application in diploma courses and Graduate level.

Poly techniques should include instrumentation and technical training so that pass- outs of these institutions can work as specialized fitters/technicians with basic and operational knowledge of fertigation.

Introduce fertigation and micro irrigation in high school syllabus.

2. Water soluble fertilizers are costly and not available in rural shops

Water soluble fertilizers need to be manufactured in the country. Import duty charged now are actually passed on to the farmers. This issue despite long discussions on several foras has not been actually attended to.

Availability of speciality fertilisers in the very rural areas are still a problem. There are no outlets in the villages keeping water soluble fertilizers. This issue was pointed out several times in the past.

3. Fertilizers for fertigation has to be included as a separate subsection and policies are to be made in such a way that the farmers are encouraged to practise fertigation

4. Fertilizer companies operating in India should take up manufacture water soluble fertilizers as a major activity. Many of them are importing and supplying now.

5. Fertigation is a stand-alone technology

It is true only thru the adoption of micro irrigation, fertigation was introduced in India. Beyond micro irrigation it has a role. The fertilizer recommendations for each crop available in University publications are several years old. This has to change. University publications should be updated taking into consideration of current information available and research results from various sources. A sincere effort on this and revision of fertilizer recommendations should be done.

6. There is a golden opportunity for Private Public participation in making fertigation a household (farmhold) process.

2

Fertigation Ratios and Scheduling

A. CEREALS

1. Rice (*Oryza sativa*)

RD 60:20:20 kg/acre NPK
(Recommended by TNAU)

Ratio

Growth Stage	Duration	N	P	K
Basal	at soil preparation	0	1	0
till 10 DAP#	10 days	0.17	0	0
11-35 DAP	25 days	0.5	0	0.17
36-55 DAP	20 days	0.17	0	0.17
56-65 DAP	10 days	0.17	0	0.5
66- 70 DAP	5 days	0	0	0.17

P is given as basal dose directly to soil

Fertilisers

Urea 130.2 kg/acre

SSP 125 kg/acre

MOP (white) 34 kg/acre

Fertigation schedule

	Urea	SSP	MOP	$ZnSO_4$	Fert Schedule (urea)	Fert Schedule MOP
Basal (soil)		125	0	10	0	0
till 10 DAP#	22.1	0	0		2.2 kg/day/acre	0
11-35 DAP	65.1	0	3.4		2.6 kg/day/acre	0.14 kg/day/acre
36-55 DAP	22.1	0	3.4		1.1 kg/day/acre	0.17 kg/day/ac
55-65 DAP	22.1	0	10		2.2 kg/day/acre	1.0 kg/day/ac
65- 70 DAP	0	0	3.4		0	0.68 kg/day/ac

DAP = Days after Planting

Micro nutrients: 10 kg/acre ZnSO4, Fertigate/or apply during 36-40 DAP

2.Wheat (*Triticum* spp.)

RD 60:24:24 kg/acre NPK (Wheat Directorate (for Haryana))

Ratio

Growth stages	Duration	N	P	K
Basal (direct to soil)	At soil preparation	0	1	0
5-20 DAG	15 days	0.35	0	0
21-80 DAG	60 days	0.65	0	0
61-120 DAG	60 days	0	0	1

DAG = Days after germination

P is applied direct to the soil as SSP at soil preparation

Fertilizers

Urea 146 kg/acre

SSP 150 kg/acre (Basal direct to soil)

MOP 40 kg/acre

ZnSO4 10 kg/acre

Fertigation Schedule

Growth stages	Duration	Urea (kg/ac)	SSP (kg/ac)	MOP (kg/ac)	Urea schedule	MOP schedule
Basal (soil)	At Soil preparation	0	150	0	0	0
5-20 DAG	15 days	48		0	3.2 kg/day/ac	0
21-80 DAG	60 days	98		0	1.6 kg/day/ac	0
61-100 DAG	60 days			40	0	1.0 kg/day/ac

3. Maize (*Zea mays*)

RD 160:115:30 NPK kg/acre

Ratio

Growth Stages	Duration	N	P	K
Basal (direct to soil)	At soil preparation	0	1	0
15-30 DAG	15 days	0.2	0	0
31-45 DAG	15 days	0.2	0	0
46-65 DAG	25 days	0.4	0	0.5
66-75 DAG	10 days	0.2	0	0.5

DAG = Days after germination

Fertilizers

Urea 250 kg/acre

DAP 250 kg/acre

MOP 50 kg/acre

Fertigation schedule

Growth stages	Duration	Urea (kg/ac)	DAP (kg/ac)	MOP (kg/ac)	Fert schedule (Urea)	Fert schedule (MOP)
Basal (soil)			250	0		
15-30 DAG	15 days	50	0	0	3.3 kg/day/ac	0
31-45 DAG	15 days	50	0	0	3.3 kg/day/ac	0
46-65 DAG	20 days	100	0	25	5 kg/day/ac	1.25 kg/day/ac
66-75 DAG	10 days	50	0	25	5 kg /day/ac for 66-70 DAG	2.5 kg/day/ac

B. PULSES

1. Pigeon pea (*Cajanus cajan*)

RD 8:20:15 kg/acre NPK

Ratios

Growth Stages	Duration	N	P	K
Basal direct to soil	At soil preparation	0	1	0
5-25 DAG	20 days	0.5	0	0
26-55 DAG	30 days	0.5	0	0.33
56-75 DAG	20 days	0	0	0.33
76-100 DAG	25 days	0	0	0.33

Fertilizers

SSP 125 kg/acre

Urea 17.4 kg/acre

MOP 25 kg/acre

Fertigation Schedule

Growth stages	Duration	Urea (kg/ac)	SSP (kg/ac)	MOP (kg/ac)	Fert schedule (Urea)	Fert schedule (MOP)
Basal (direct to soil)	At Soil Preparation	0	125	0		0
5-25 DAG	20 days	10		0	0.5 kg/day/ac	0
26-55 DAG	30 days	10		8.5	0.5 kg/day/ac	0.5 kg/day/ac
56-75 DAG	20 days	0		8.5	0	0.5 kg/day/ac
76-100 DAG	25 days	0		8.5	0	0.5 kg/day/ac

1. Rhizobium to be applied on the day of planting seed
2. SSP to be applied and incorporated during the last plough operation

2. Blackgram (*Vigna mungo*)

RD 10:10:10 kg/acre NPK

Ratios

Growth Stages	Duration	N	P	K
Basal	At Soil preparation	0	1	0
1-35 DAE	35 days	0.55	0	0
36-56 DAE	20 days	0.45	0	0.3
57-75 DAE	20 days	0	0	0.7

Fertigation Schedule $$

Growth stages	Duration	Urea (kg/ac)	SSP (kg/ac)	MOP (kg/ac)	Fert schedule (Urea)	Fert schedule (MOP)
Basal (soil) At Planting	At Soil Preparation	0	62.5	0		
1-35 DAE		12		0	1kg/3 days/ac	0
36-55 DAE		9.7		4.5	1.4 kg/3 days/ac	0.6 kg/3 days/ac
57-75 DAE		0		12.2	0	2 kg/3 days/ac#

P is applied direct to soil at land preparation as SSP and it will provide Sulphur (12.5 kg/acre)
#Last MOP dose is 2.2 kg

C. OIL SEEDS CROPS (Seasonal)

1. Soybean (*Glycine max*)

RD 12:24:16 kg/acre NPK

Ratios

Growth stages	Duration	N	P	K
Basal	At Soil Preparation	0	1	0
1-35 DAE	35 days	0.5	0	0
36-56 DAE	20 days	0.5	0	0.5
57-90 DAE	33 days	0	0	0.5

Fertigation schedule

	Duration	Urea (kg/ac)	SSP (kg/ac)	MOP (kg/ac)	Schedule of Urea	Schedule of SSP	Schedule of MOP
Basal at Planting	Soil incorporation	250#	0	0	full as basal	0	
1-35 DAE	35 days	13.02	0	0	1.2 kg/3 days (11 Doses)	0	0
36-56 DAE	20 days	13.02	0	13.4	1.9 kg/3 days (7 doses)		1.9 kg/3days (7 doses)
57-90 DAE	33 days	0	0	13.4	1.2 kg/3 days (11 doses)		1.2 kg/3days (7 doses)

SSP supplies Sulphur also (50 S)

2. Sunflower (*Helianthus annuus*)

RD 24:36:16 kg/acre NPK

Ratios

Growth stages	Duration	N	P	K
Basal	At Soil preparation	0	1	0
5 -30 DAE	25 days	0.6	0	0
31-45 DAE	15 days	0.4		0.25
46-60 DAE	15 days	0	0	0.5
61-70 DAE	10 days	0	0	0.25

Fertigation schedule

Growth Stages	Duration	Urea	SSP	MOP white	Schedule of Urea	Schedule of MOP
Basal direct to soil	At Soil Preparation	0	225	0		
5 -30 DAE	25 days	31.24	0		3.9 kg/ac/3 days (8 doses)	0
31-45 DAE	15 days	20.8		6.68	4.2 kg/ac/3 days (5 doses)	1.3 kg/ac/3 days (5 doses)
46-60 DAE	15 days	0	0	13.36	0	2.7 kg/ac/3 days (5 doses)
61-70 DAE	10 days	0	0	6.68	0	2.2 kg/ac/3 days (3 doses)

SSP as P source will also supply Sulphur essential for Sunflower

3. Castor *Ricinus communis* (Kharif crop; June-September)

Duration 120 days

RD 24:24:16 kg/acre NPK (Andhra Pradesh & Telengana)

Ratios

Growth stages	Duration	N	P	K
Basal direct to soil	At Soil preparation	0	1	0
5 DAE- 30 DAE	25 days	0.2	0	0
31-75	45 days	0.4	0	0.2
75-90	15 days	0.4	0	0.5
91-110	20 days	0	0	0.3

Fertigation schedule

Growth Stages	Duration	Urea	SSP	MOP white	Schedule of Urea	Schedule of MOP
Basal direct to soil	At Soil Preparation	0	150	0		
5-30 DAE	25 days	15.6	0	0	2 kg/ac/3 days (8 doses)	0
31-75 DAE	45 days	26	0	5.3	1.74 kg/ac/3 days (15 doses)	0.35 kg/ac/3 days (15 doses)
76-90 DAE	15 days	10.4	0	13.4	2 kg/ac/3 days (5 doses)	2.7 kg/ac/3 days (5 doses)
91-110 DAE	20 days	0	0	8	0	1.3 kg/ac/3 days (6 doses)

(sulphur 12 kg/acre). SSP applied for P contains Sulphur

Micronutrients- 5 kg $ZnSO_4$ /acre Fertigate on 31 -75 DAE
10 kg $FeSO_4$/acre AS soil application

3a. Castor – Rabi Season (October- February)

Rabi Castor 160 day duration

RD 48:24:16 kg/acre NPK

Ratios

Growth Stages	Duration	N	P	K
Basal direct to soil	At Soil Preparation	0	1	0
5 DAE- 30 DAE	25 days	0.2	0	0
31-60	30 days	0.3	0	0.1
61-105	45 days	0.4	0	0.2
106-130	25 days	0.1	0	0.3
131-150	20 days	0	0	0.4

Fertigation Schedule

Growth Stages	Duration	Urea	SSP	MOP white	Schedule of Urea	Schedule of MOP
Basal direct to soil	At Soil Preparation		150			0
5-30 DAE	25 days	20.84	0	0	2.6 kg/ac/3 days (8 doses)	0
31-60 DAE	30 days	31.26	0	2.7	2.08 kg/ac/3 days (15 doses)	0.35 kg/ac/3 days (15 doses)
61-105 DAE	45 days	41.68	0	5.3	8.3 kg/ac/3 days (5 doses)	2.7 kg/ac/3 days (5 doses)
106-130 DAE	25 days	10.42	0	8.1	1.7 kg/ac/3 days (6 doses)	1.35 kg/ac/3 days (6 doses)
131-150 DAE	20 days	0	0	10.7	0	1.8 kg/ac/3 days (6 doses)

Micronutrients	5 kg ZnSO4 /acre	Fertigate on 31 -75 DAE
	10 kg FeSO4/acre	AS soil application

4. Mustard (*Brassica rapa*)

RD 48:24:24 kg/acre NPK

Ratios

Growth Stages	Duration	N	P	K	Mg SO_4
Basal direct to soil	At Soil Preparation	0	0.5	0	
Germination- 10 DAE		0.05	0	0	
11-30 DAE		0.4	0.2	0	
31-65 DAE		0.4	0.3	0.2	10 kg/ha
66-90 DAE		0.15	0	0.5	10 kg/ha
91-100 DAE		0	0	0.3	

Sulphur required for oil comes from Pot. sulphate and the basal dose of SSP.

Fertigation Schedule

Growth Stages	Duration	Urea (kg/ac)	SSP (kg/ac)	MAP (Kg/ac)	SOP (kg/ac)	Schedule of Urea	Schedule of MAP	Schedule of SOP
Basal (direct to soil)	At Soil Pre-paration	0	75	0	0		0	0
Germin-ation-10 DAE	10 days	5	0	0	0	1.7 kg/3 days/ac (3 doses)	0	0
11-30 DAE	20 days	40	0	7.9	0	6.7 kg/3 days/ac (6 doses)	1.3 kg/3 days/ac (6 doses)	0
31-65 DAE	35 days	40	0	11.8	9.6	3.3 kg/3 days/ac (12 doses)	1.2 kg/3 days/ac (10 doses)	0.8 kg/3 days/ac (12 doses)
66-90 DAE	25 days	15	0	0	24	3 kg/3 days/ac (5 doses)	0	2 kg/3 days/ac (12 doses)
91-100 DAE	10 days	0	0	0	14.4	0	0	4.8 kg/3 days/ac (3 doses)

5. Safflower (*Cartha mustinctorius*)

RD 16:10:10 kg/acre NPK (Directorate of Oil Seed Research)

Ratios

Growth Stages	Duration	N	P	K
Basal (direct to soil)	At Soil Preparation	0	1	0
10 -39 DAE	30 days	0.5	0	0
40- 60 DAE	20 days	0.3	0	0.3
61-72 DAE	12 days	0.2	0	0.2
73-90 DAE	18 days	0	0	0.5

Fertigation schedule

Growth Stages	Duration	Urea	SSP	MOP white	Schedule of Urea	Schedule of MOP
Basal (direct to soil)	At Soil Preparation	0	62.5	0		
10 -39 DAE	30 days	31.5		0	3.15 kg/3 days/ ac (10 doses)	
40-60 DAE	20 days	18.8		5.01	2.7 kg/3 days/ac (7 doses)	0.72 kg/3 days/ ac (7 doses)
61-72DAE	12 days	12.5		3.3	3.1kg/3 days/ac (4 doses)	0.83 kg/3 days/ ac (4 doses)
73-90 DAE	18 days	0		8.4	0	1.4 kg/3 days/ac (6 doses)

6. Ground nut (*Arachishy pogaea*)

RD 6.6:40:90 kg/acre NPK

Soil application

SSP 250 kg/ac

Muriate of Potash 75 kg/ac

Fe-EDTA 1.5 kg/ac (along with FYM @ 5t/ac)

Gypsum 2 ton/ac Soil application

Fertigation

Ammonium Sulphate 10 kg/ac

Calcium Nitrate 30 kg/ac

Fe –EDTA 1.5 kg/ac

Muriate of Potash 75 kg/ac

Fertigation Schedule

Growth Stages	Duration	Ammo. Sulphate	Calcium Nitrate	SSP	MOP white	Fe-EDTA	Schedule of Ammo. Sulphate	Schedule of Calcium Nitrate	Schedule of MOP	Schedule of Fe-EDTA
Basal (Direct to soil)	At soil pre-paration			250	75	1.5				
7-21 DAE	14 days	10	0		15		2 kg/ac/3 days (5 doses)	0	3 kg/ac/3 days (5 doses)	0.5 kg/ac on 10 DAE
22-65 DAE		43 days	0	30		60	0	3 kg/ac/3 days (10 doses)	5 kg/ac/3 days (12 doses)	0.5 kg/ac each on 25 and 45 DAE

Gypsum @ 2 t/acre for enriching Ca and promoting gynophore growth.

D. OIL SEEDS CROPS (Perennial)

1. Coconut (*Cocosnu cifera*)

RD (Southern Coastal region, India)

	Fertilizer recommendation for Coconut				
	N g/tree	**P g/tree**	**K g/tree**	**OM**	
yr1	Planting in May June			12.5 t/ha	Apply before planting
yr1	50	40	135		in September-October
yr2	50	40	135	25 kg/tree	in June-July
	110	80	270	25 kg/tree	in September-October
yr3	110	80	270	25 kg/tree	in June-July
	220	160	540	25 kg/tree	in September-October
yr4	170	120	400	25kg/tree	in June-July
Plus	330	200	800	25 kg/tree	in September-October

Lime application for Coastal area

2 kg/tree per year up to 15 years

4 kg/tree per year after 15 years

Lime should be incorporated 15 days prior to fertilizer application in September

Micronutrient/Secondary nutrient

Apply 0.5kg/tree Magnesium Sulphate for bearing trees

Year 1 Palm

RD NPK 50:40:135 g/tree

Fertigation schedule

Fertilizer	Rate g/tree	Fertigation rate	Duration	
Urea	110 g	20 g/tree/week	Sept 1wk to Oct 2wk	6 doses
MOP	225 g	40 g/tree/week	Sept 1wk to Oct 2wk	6 doses

Phosphate, Apply 250 g SSP in the form of a ring around the tree 50 cm away from plant and 15 cm deep in September first week.

Year 2 palm

RD NPK 50:40:135 g/tree (June- July) and 110:80:270 g/tree (Sept-Oct)

Fertigation schedule

Fert type	Rate g/tree	Fertig rate	Duration	
Urea	110 g	14 g/tree/wk	June 1 wk to July 4 wk	8 doses
MOP	225 g	28 g/tree/wk	June 1 wk to July 4 wk	8 doses
Urea	239 g	30 g/tree/wk	Sept 1wk to Oct 4wk	8 doses
MOP	450 g	56 g/tree/wk	Sept 1wk to Oct 4wk	8 doses
Phosphate ; SSP	250 g/tree			

Apply 250 g SSP in the form of a ring around the tree 50 cm away from plant and 15 cm deep in June first week.

Apply 500 g SSP in the form of a ring around the tree 50 cm away from plant and 15 cm deep in September first week.

Year 3 palm

RD 110:80:270 g/tree (June July) and
220:160:540 g/tree (September –October)

Fertigation Schedule

Fertilizer	Rate g/tree	Fertigation rate	Duration	
Urea	239 g	30 g/tree/wk	June 1 wk to July 4 wk	8 doses
MOP	450 g	56 g/tree/wk	June 1 wk to July 4 wk	8 doses
Urea	478 g	60 g/tree /week	Sept 1wk to Oct 4wk	8 doses
MOP	900 g	112 g/tree/week	Sept 1wk to Oct 4wk	8 doses
Phosphate;SSP	500 g/tree			

Apply 500 g/tree SSP in the form of a ring around the tree 50 cm away from trunk and 15 cm deep in first week of June

Apply SSP in the form of a ring around the tree 50 cm away from the trunk and 15 cm deep.

500 g/tree in 1st week of September and 500 g/tree in Last week of October

Year 4 plus aged palm

RD NPK 170:140:400 g/tree (Jun-July) and 330:200:800 g/tree (Sept-Oct)

Fertigation schedule

Fertilizer	Rate g/tree	Fertigation rate	Duration	
Urea	370 g	13.2 g/tree/2 days	June 1wk to July 4 wk	30 doses
MOP	665 g	22.2g/tree/2 days	June 1wk to July 4 wk	30 doses
Urea	718 g	24 g/tree/2 days	Sept 1 wk to Oct 4 wk	30 doses
MOP	1335 g	44.5 g/tree/2 days	Sept 1 wk to Oct 4 wk	30 doses

Apply SSP in the form of a ring around the tree 1 m away from trunk and 15 cm deep.

First application 750 g/tree in June first week and second application 1250 g/tree in September first week.

Fertigation for all nutrients, N,K and P; with Ammonium sulphate (for N) and Phosphoric acid (for P).

Year 4 palm

RD 170:120:400 g/tree in June_ July

330:200:800 g/tree in September –October

Fertigation schedule

Fertilizer	Rate g/tree	Fertigation rate	Duration	
Ammonium sulphate	810 g/tree	27g/tree/2 days	June 1wk to July 4 wk	30 doses
Phosphoric acid	200 g/tree	6.7 g/tree/2 days	June 1wk to July 4 wk	30 doses
MOP	665 g/tree	22.2 g/tree/2 days	June 1wk to July 4 wk	30 doses
Ammo sulphate	1571 g/tree	52 g/tree/2 days	Sept 1 wk to Oct 4 wk	30 doses
Phosphoric Acid	333 g/tree	11.1 g/tree/2 days	Sept 1 wk to Oct 4 wk	30 doses
MOP	1333 g/tree	44.4 g/tree/2 days	Sept 1 wk to Oct 4 wk	30 doses

2. Oil Palm (*Elaeis guineensis*)

Year 1 (28 days after transplanting to week 52; week 5 to 52)

RD 400 N: 200 P: 900 K g/palm/year

Fertilizers

Urea 868 g/tree

SSP 1250 g/tree
(Apply direct to soil)

MOP 1494 g/tree

Fertigation Schedule

Fertigation Schedule - one dose per week	Dose of Urea	Dose of MOP
Duration	Urea g/palm	MOP g/palm
Day 1- Day 28 (first 4 weeks)	0	0
Week 5-52 (48 doses)	18 g/dose (48 doses)	31 g/dose (48 doses)

Year 2 (week 1 to 52)

RD 800 N: 400 P: 1800 K g/palm/year

Fertilizers

Urea 1736 g/tree

SSP 2500 g/tree (Apply direct to soil)

MOP 2988 g/tree

Fertigation schedule

Fertigation Schedule - one dose per week	Dose of Urea	Dose of MOP
Duration	Urea g/palm	MOP g/palm
Week 1-52 (52 doses)	36 g/dose (52 doses)	62.5 g/dose (52 doses)

Year 3 (week 1 to 52)

RD 1200 N: 600 P: 2700 K g/palm/year

Fertilizers

Urea 2604 g/tree

SSP 3750 g/tree (Apply direct to soil)

MOP 4482 g/tree

Fertigation Schedule

Fertigation Schedule - one dose per week	Dose of Urea	Dose of MOP
Duration	Urea g/palm	MOP g/palm
Week 1-52 (52 doses)	54.5 g/dose (52 doses)	93.5 g/dose (52 Doses)

P application each year
SSP should be ring applied in 3 splits directly to the soil.
10-15 cm deep furrow at 1 m away from trunk. Apply fertilizer in the furrow and cover with soil.
For 5 plus year tree make two rings 1m, and 2m away from the trunk

E. VEGETABLES

1. Potato (*Solanum tuberosum*)

Potato is both a direct food and a vegetable in the country.

RD 50:25:50 NPK kg/acre

Total Fertilizer Requirement

Fertilizer	Kg/acre
DAP	54.3
Ammonium sulphate	100
Urea	41.7
MOP	83.1

Ratio

Duration	N	P	K
Basal at Planting	0.2	1	0
15-30 days	0.12	0	0
31-55 days	0.57	0	0.45
56-70 days	0.11	0	0.26
71-90 days	0	0	0.29
	1	1	1

Fertigation schedule

Duration	Fertilizer	Kg/acre	Kg/ac/day	No of doses
Basal Soil application	DAP	54.3		1
15-30 days	Ammonium sulphate	28.5	1.9	15
31-45 days	Ammonium sulphate	71.5	4.77	15
46-55 days	Urea	29.3	2.93	10
31-55 days	MOP	37.5	1.5	25
56-70 days	Urea	12.4	0.82	15
	MOP	21.6	1.4	15
71-90 days	MOP	24	1.25	19

2. Tomato (*Lycopersicumes culentum*)

RD 80:100:100 kg/ac NPK

Ratio

	N	P	K
stage 1	0.1	0.2	0.05
stage 2	0.4	0.4	0.4
stage 3	0.3	0.2	0.3
stage 4	0.2	0.2	0.25

Stage of crop	Duration	Fertilizer	Total Quantity kg/ac	Dose
1. Plant establishment (4 -14 DAP)	10 days	19:19:19	26.5	2.65 kg/day/acre
		13:00:46	11	1.1 kg/day/acre
		Urea	3.5	1.2 kg every 3 days
2. Flower initiation stage (15 -45 DAP)	30 days	0.542361111	16.4	1.6 kg every 3 days
		13:00:46	89	8.9 kg every 3 days
		Urea	40	4 kg every 3 days
3. Flower to fruit set (46-75DAP)	30 days	19:19:19	26.5	2.6 kg every 3 days
		13:00:46	55.5	5.5 kg every 3 days
		Urea	25.5	2.5 kg every 3 days
4. fruit set and maturity (76-145 DAP)	70 days plus	0.542361111	8	1 kg every 3 days
		13:00:46	44.5	3.2 kg every 3 days
		Urea	20	3 kg every 3 days

DAP - Days after planting

3. Brinjal (*Solanumme longena*)

RD NPK 50:40:20 kg/acre

Fertilizers: Urea 108.5 kg/acre

Phosphoric acid 76.8 kg/acre

White MOP 33.4 kg /acre

Ratios

Growth stages	Duration (Days)	N	P	K
Initial	0-10	0.3	0.25	0.1
Vegetative	11 to 30	0.5	0.5	0.15
Flowering/fruit	31to 60	0.2	0.25	00.25
Maturity	61 to 100	0	0	0.5

Fertigation Schedule

Stage of crop	Duration	Fertilizer	Total Quantity kg/ac	Dose
1. Plant establishment (4 -10 DAP)	6 days	Urea	21.7	3.62 kg/day/acre
		Phosphoric acid	19.2	3.2 kg/day/acre
		MOP	0	0
2. Vegetative (11-30 DAP)	20 days	Urea	54.4	8.12 kg every 3 days
		Phosphoric acid	38.4	5.76 kg every 3 days
		White MOP	5	0.75 kg every 3 days
3. Flower to fruit set (31-60 DAP)	30 days	Urea	32.6	3.27 kg every 3 days
		Phosphoric acid	19.2	1.92 kg every 3 days
		White MOP	11.7	1.17 kg every 3 days
4. Fruit set and maturity (61-100 DAP)	40 days	Urea	0	0
		Phosphoric acid	0	0
		White MOP	16.7	1.2 kg every 3 days

DAP = Days after Planting (transplanting)

4. Beans (*Phaseolus* spp.)

RD 80:88:108 NPK kg/acre

Fertilizers

Fertilizers	Total Quantity
Urea	118 kg (54.35 N)
19:19:19	100 kg (19 N, 19 P, 19 K)
SSP	220 kg (35.2P)
MAP	55.4 kg (6.65 N, 33.8 P)
SOP	216 kg (108 K)

Ratio

Stages		N	P	K
At Land preparation	Basal	0	0.4	0
5-30 DAE		0.4	0.3	0.2
31- 75 DAE		0.4	0.3	0.4
76-90 DAE		0.2	0	0.4

Fertigation schedule

Stage of crop	Duration	Fertilizer	Total Quantity kg/ac	Dose
At Land preparation		SSP	220	Full at land preparation
5-30 DAE	25 Days	Urea (16-30DAE)	25.1	1.7 kg/day/acre
		MAP (16-30 DAE)	12.1	0.81 kg/day/acre
		19:19:19 (5-30 DAE)	100	4 kg/day/acre
		SOP (16-30 DAE)	43.2	2.9 kg/day/acre
31- 75 DAE	45 days	MAP	43.3	2.9 kg/3 days/acre
		Urea	69.2	4.61 kg/3 days/acre
		SOP	129.6	8.64 kg/3 days/acre
76-90 DAE	15 Days	Urea	23.6	4.72 kg/3 days/acre
		SOP	43.2	8.64 kg/3 days/acre

5. Okra (lady's finger) (*Abelmoschuses culentus*)

RD 80:40:45 kg/acre NPK.
For South Karnataka (Mandya)

Ratios

		N	P	K
Basal	At bed preparation	0.14	0.7	0
	10-20 DAE	0.1	0	0
	21-50 DAE	0.3	0.1	0.3
	51-75 DAE	0.3	0.2	0.4
	76-100 DAE	0.16	0	0.3

Fertilizers

	kg/acre	N	P	K
DAP	**60**	10.8	28	0
MAP	**20**	2.4	12	0
Urea	**145**	66.8	0	0
SOP	**90**	0	0	45
		80	**40**	**45**

Fertigation schedule

Stages	Duration	Fertilizer	Total Quantity kg/acre	Dose
At Land preparation		DAP	60	Full at land preparation
10-20 DAE	10 days	Urea	14.5	1.45 kg/day/acre (10 doses)
21- 50 DAE	30 days	MAP	6.6	0.67 kg/3 days/acre (10 doses)
		Urea	50.4	5.04 kg/3 days/acre (10 doses)
		SOP	27	2.7 kg/3 days/acre (10 doses)
51-75 DAE	25 days	MAP	13.4	1.68 kg/3 days/acre (8 doses)
		Urea	48.6	6.1 kg/3 days/acre (8 doses)
		SOP	36	4.5 kg/3 days/acre (8 doses)
76-90 DAE	15 days	Urea	27.8	5.6 kg/3 days/acre (5 doses)
		SOP	27	5.4 kg/3 days/acre (5 doses)

6. Cucumber (open field) (*Cucumis sativus*)

RD NPK 60:30:30 kg/acre (TNAU)

Ratios

	N	P	K
Basal 50 kg DAP (9N 23 P)	0.15	0.77	
11-20 DAE	0.1		0
21-40 DAE	0.35	0.23	0.2
41-80 DAE	0.4		0.5
81-95 DAE			0.3

Fertilizers

DAP basal	50 kg /ac	N 9	P 23
MAP for balance P	11.5 kg/ac	1.4 N	P 7
Urea for balance N 60- (9+1.4) 10.4	107.6 kg/ac	49.6 N	0
White MOP	50 kg /ac	30 K	

Fertigation schedule

Stages	Duration	Fertilizer	Total Quantity kg/acre	Dose
At Land preparation		DAP	50	Full at land preparation
11-20 DAE	10 days	Urea	13	1.3 kg/day/acre (10 doses)
21- 50 DAE	30 days	MAP	11.5	1.15 kg/3 days/acre (10 doses)
		Urea	45.57	4.6 kg/3 days/acre (10 doses)
		White MOP	10	1.0 kg/3 days/acre (10 doses)
51-80 DAE	30 days	Urea	52.1	5.21 kg/3 days/acre (10 doses)
		White MOP	25.1	2.51 kg/3 days/acre (10 doses)
81-95 DAE	15 days	White MOP	15	3 kg/3 days/acre (5 doses)

7. Gherkins (*Cucumis* spp.)

RD 95:63:245 NPK kg/acre (Gherkin contract farming Company GGCL)

Ratio

	N	P	K
Basal	0.2	0.7	0.12
10-20 DAE	0.2	0	0.18
21-39 DAE	0.3	0.3	0.5
40-65 DAE	0.3	0	0.2

Fertilizers

DAP (Diammonium Phosphate);

CAN (Calcium Ammonium Phosphate); 19:19:19;

White MOP and MgSO4 (Magnesium Sulphate)

Fertigation schedule

Stages	Duration	Fertilizer	Total Quantity kg/acre	Dose
At Land preparation		DAP	50	Full at land preparation
		MOP	50	Plus MgSO4 20 kg
10-20 DAE	10 days	CAN	73.15	7.32 kg/day/acre (10 doses)
		MOP	73.65	7.37 kg/day/acre (10 doses)
21-40 DAE	20 days	CAN	0	0
		19:19:19	99.4	4.97 kg/day/acre (20 doses)
		White MOP	173	8.65 kg/day/acre (20 doses)
41-65 DAE	25 days	CAN	109.7	4.4 kg/day/acre (25 doses)
		White MOP	81.83	3.27 kg/day /acre (25 doses)

8. Cauliflower (*Brassica oleracea var. botrytis*)

RD 100:60:60 NPK kg/ac (for UP)

Ratio

	N	P	K
Basal	0.1	0.4	0
10-20 DAT	0.3	0.2	0.1
21-35	0.5	0.4	0.3
36-50	0.1		0.4
51-70			0.2

Fertigation schedule

Stages	Duration	Fertilizer	Total Quantity kg/acre	Dose
At Land preparation		DAP	50	Full Soil Application
10-20 DAT	10 days	MAP	20.2	2.02 kg/day/acre (10 doses)
		Urea	60	6 kg/day/acre (10 doses)
		White MOP	10	1 kg/day/acre (10 doses)
21-35 DAT	15 days	MAP	40.5	2.7 kg/day/acre (15 doses)
		Urea	98	6.5 kg/day/acre (15 doses)
		White MOP	30	2 kg/day/acre (15 doses)
36-50 DAT	15 days	Urea	22	3.1 kg/2 days/acre (7 doses)
		White MOP	40	5.7 kg/2 days/acre (7 doses)
51-60 DAT	10 days	White MOP	20	4 kg/2days/acre (5 doses)
61-70 DAT	10 days	0	0	0

Fertigation begins at 10 days after transplanting (DAT)

9. Cabbage (*Brassica oleracea var. capitate*)

RD 80:50:60 kg/ac NPK (TN)

Ratios

	N	P	K
Basal	0.11	0.46	0
10-40 DAT	0.5	0.34	0.1
41-70	0.3	0.2	0.5
71-90	0.09		0.4
91-105	0	0	0

Basal 50 kg DAP (9 N and 23 P)

Fertigation schedule

Stages	Duration	Fertilizer	Total Quantity kg/acre	Dose
At Land preparation		DAP	50	Full Soil Application
10-40 DAT	30 days	MAP	30	1 kg/day/acre (30 doses)
		Urea	75	2.5 kg/day/acre (30 doses)
		White MOP	10	1 kg/day/acre (from 31-40 DAT) (10 doses)
41-70 DAT	30 days	MAP	14.5	0.97 kg/ 2 days/acre (15 doses)
		Urea	50	3.3 kg/ 2 days/acre (15 doses)
		White MOP	50.1	1.67 kg/ 2 days/acre (15 doses)
71-90 DAT	30 days	Urea	17.8	1.2 kg/2 days/acre (15 doses)
		White MOP	24	1.6 kg/2 days/acre (15 doses)
91-105 DAT		0	0	0

Basal 50 kg DAP
Fertigation begins at 10 days after transplanting (DAT)

10. Onion (*Allium cepa*)

RD 60:50:80 kg/ac NPK (TNAU)

Fertilizers

SSP 150 kg/ac Apply direct to soil

Urea 118.7 kg/ac;

MAP 42.64 kg/ac;

MOP 133.6 kg/ac

Ratio

Growth period	N	P	K
Basal 150 kg/ac SSP	0	0.48	0
3-15 DAG	0.15	0	0
16-45 DAG	0.5	0.26	0.25
46-60 DAG	0.35	0.26	0.25
61-75 DAG	0	0	0.5

Fertigation Schedule

Stages	Duration	Fertilizer	Total Quantity kg/acre	Dose
At Land preparation		SSP	150	Full Soil Application
3-15 DAG	12 days	Urea	19.3	3.2 kg/2 days/acre (6 doses)
16-45 DAG	30 days	MAP	21.3	1.42 kg/ 2 days/acre (15 doses)
		Urea	72.56	4.83 kg/ 2 days/acre (15 doses)
		White MOP	33.4	2.2 kg/ 2 days/acre (15 doses)
46-60 DAG	15 days	MAP	21.3	3.0 kg/2 days/acre (7 doses)
		Urea	27	3.85 kg/2 days/acre (7 doses)
		White MOP	33.4	4.8 kg/2 days/acre (7 doses)
61-75 DAG	15 days	White MOP	66.8	9.54 kg/2 days/acre (7 doses)

11. Ridge (Ribbed) gourd (*Luffaacu tangula*)

12. Bitter gourd (same schedule for both gourd vegetables) (*Momordica charantia*)

Ridge Gourd

Bitter Gourd

RD 100:40:40 kg/acre NPK for both (TNAU)

Ratios

Ratio	Duration	N	P	K
Basal	only OM			
5 -10 DAE	5 days	0.1	0.2	0
11-40 DAE	30 days	0.2	0.4	0.1
41-60 DAE	20 days	0.3	0.3	0.3
61-110 DAE	50 days	0.4	0.1	0.6
111-120 DAE	10 days	0	0	0

Fertilizers

Urea

MAP

Potassium Nitrate

Fertigation schedule

Stages	Duration	Fertilizer	Total Quantity kg/acre	Dose
At Land preparation		Only Organic Manure		Full Soil Application
5-10 DAE	5 days	MAP	12.9	2.6 kg/day/acre (5 doses)
		Urea	18.4	3.7 kg/day/acre (5 doses)
11-40 DAE	30 days	MAP	25.8	1.72 kg/ 2 days/acre (15 doses)
		Urea	34.3	2.29 kg/ 2 days/acre (15 doses)
		Potassium Nitrate (from 20 DAE)	8.7	0.87 kg/ 2 days/acre (10 doses)
41-60 DAE	20 days	MAP	19.7	1.97 kg/2 days/acre (10 doses)
		Urea	52.5	5.25 kg/2 days/acre (10 doses)
		Potassium Nitrate	26	2.6 kg/2 days/acre (10 doses)
61-110 DAE	50 days	MAP (till 80 DAE)	6.6	0.66 kg/2 days/acre (10 doses)
		Urea	70.3	2.81 kg/2 days/acre (25 doses)
		Potassium Nitrate	52.1	2.1 kg/2 days/acre (25 doses)
111-120 DAE	10 days	nil	0	0

Sulphate and Chloride fertilizers are not suitable for Gourd vegetables. Hence, MAP, Urea, and Potassium nitrate are chosen.

13. Moringa (*Moringao leifera*)

RD 50:50:50 NPK kg/acre (APAU)

Ratios

Growth period	N	P	K
Basal 155kg/ac SSP	0	0.5	0
10-30 DAP	0.2	0	0
31-90 DAP	0.5	0.5	0.3
91-180 DAP	0.3		0.7

Basal dose applied to the planting pit (10kg FYM+ 230 g SSP+ 250 g Neem cake +30 g MgSO4./pit

155 kg/acre SSP (25 P)

Fertilizers

Ammonium sulphate 214.8 kg/acre;

MAP 41 kg/acre

White MOP 83.5 kg/acre

Fertigation schedule##

Stages	Duration	Fertilizer	Total Quantity kg/acre	Dose
Basal 155 kg/ac SSP				
10-30 DAP	20 days	Ammonium Sulphate	47.6	4.76 kg/2 days/acre (10 doses)
		MAP	0	0
		White MOP	0	0
31-90 DAP	60 days	Ammonium Sulphate	95.2	3.17 kg/2 days/acre (30 doses)
		MAP	41	1.37 kg/2 days/acre (30 doses)
		White MOP (from 50 DAP)	25.1	1.25 kg/2 days/acre (20 doses)
91-180 DAP	90 days	Ammonium Sulphate (till 150 DAP)	71.4	2.38 kg/2 days/acre (30 doses)
		MAP	0	0
		White MOP	58.5	1.3 kg/2 days/acre (45 doses)

Repeat the schedule a week after pruning for the next cycle

14. Radish (*Raphanus sativus*)

RD 25:12:12 kg/acre NPK (Punjab)

Ratio

	N	P	K
Basal 21 kg DAP	0.4	0.3	0
5-15 DAE	0.2	0.4	0.1
16-30 DAE	0.3	0.3	0.5
31-50 DAE	0.1	0	0.4

Fertilizers

DAP 21 kg	10	4	0
MAP 13 kg	1.6	8	0
Urea 29 kg	13.4	0	0
SOP 24 kg	0	0	12

Ratio

Growth period	Duration	Fertilizer	Total Quantity kg/acre	Dose
Basal 21 kg/ac DAP		DAP	21	At soil preparation
5-15 DAE	10 days	Urea	8.9	1.78 kg/2 days/acre (5 doses)
		MAP	7.4	1.48 kg/2 days/acre (5 doses)
		SOP (from 12 DAE)	2.4	1.2 kg/2 days/acre (2 doses)
16-30 DAE	15 days	Urea	14.82	2.11 kg/2 days/acre (7 doses)
		MAP	5.6	0.8 kg/2 days/acre (7 doses)
		SOP	12	1.71 kg/2 days/acre (7 doses)
31-50 DAE	20 days	Urea	5.42	0.54 kg/2 days/acre (10 doses)
		MAP	0	0
		SOP	9.6	0.96 kg/2 days/acre (10 doses)

15. Carrot (*Daucu scarota*)

RD 50:40:120 kg/acre NPK (South Karnataka)

Ratios

Growth periods	Duration (days)	N	P	K
Basal	Basal dose at Bed formation	0.1	0.3	0
5-15 DAE	10	0.3	0.4	0
16-35 DAE	20	0.4	0.3	0.15
36-60 DAE	30	0.2	0	0.5
61-90 DAE	30	0	0	0.35

Fertilizers

Basal

DAP

$CaSO_4$

$MgSO_4$

Fertigation

Urea

Phosphoric Acid

Potassium Sulphate (SOP)

Fertigation Schedule

Growth period	Duration	Fertilizer	Total Quantity kg/acre	Dose
Basal	At land preparation and Bed making	DAP $CaSO_4$ $MgSO_4$	28.7 200 20	full full part (10 kg later in Fertigation)
5-15 DAE	10 days	Urea Phosphoric Acid SOP	32.6 30.72 0	3.3 kg/ day/acre (10 doses) 3.72 kg/ day/acre (10 doses) 0

16-35 DAE	20 days	$MgSO_4$	10	1 kg/2 days/acre (10 doses)
		Urea	43.4	4.34 kg/2 days/acre (10 doses)
		Phosphoric Acid	23	2.3 kg/2 days/acre (10 doses)
		SOP	36	3.6 kg/2 days/acre (10 doses)
36-60 DAE	30 days	Urea	21.7	1.4 kg/2 days/acre (15 doses)
		Phosphoric Acid	0	0
		SOP	120	8 kg/2 days/acre (15 doses)
61-90 DAE	30 days	Urea	0	0
		Phosphoric Acid	0	0
		SOP	84	5.6 kg/2 days/acre (15 doses)

Chloride toxicity will harm the crop. Therefore SOP is chosen as source for K.

19. Capsicum (*Capsicum annuum*) (open field)

RD 100:60:60 kg/acre NPK

Ratio

Crop Stage	Duration	N	P	K
Basal	Land preparation	0	0.6	0
5-15 DAP	10	0.1	0.08	0.08
16-45 DAP	30	0.3	0.16	0.17
46-75 DAP	30	0.2	0.08	0.25
75-170 DAP	95	0.4	0.08	0.5

Fertilizers

Single Super Phosphate (SSP)

Water Soluble NPK (19:19:19)

Potassium nitrate

Mono Ammonium Phosphate

Fertigation Schedule

Stages of crop	Duration	Fertilizer	Total Quantity kg/acre	Schedule
Basal	Land preparation	SSP	218	All applied at bed preparation
5-15 DAP	10 days			
		19:19:19	26.5	5.3 kg/2 days/acre (5 doses)
		Urea	11	2.2kg /2 days/acre (5 doses)
16-45 DAP	30 days	Urea	55	3.7 kg/2 days/acre (15 doses)
		0.542361	16.4	1.1 kg/2 days/acre (15 doses)
		13:00:46	22	1.5 kg/2 days/acre (15 doses)
46-75 DAP	30 days	Urea	26	1.7 kg / 2 days/acre (15 doses)
		19:19:19	26.5	1.8 kg / 2 days/acre (15 doses)
		13:00:46	22	1.5 kg /2 days/acre (15 doses)
76-165 DAP	90 days	Urea	66	1.5 kg/2 days/acre (45 doses)
		0.542361	8	0.5 kg/ 2 days /acre (16 doses)
		13:00:46	66	1.5 kg/2 days/acre (45 doses)

F. SPICE CROPS

1. Aijwan (*Trachysper mumammi*)

RD 40: 20 :20 Kg/acre NPK

Ratio

	N	P	K
Basal	0	0.5	0
10-45 days	0.4	0.25	0
46-90 days	0.4	0.25	0
91-140 days	0.2	0	1

Fertilizers

- SSP (apply direct to soil)
- Urea
- MAP
- MOP (white)

Fertigation Schedule

Growth stage	Duration	SSP kg/ac	MAP kg/ac	MAP Dose	Urea kg/ac	Urea Dose	White MOP kg/ac	White MOP dose
Basal	at land preparation	62.5 kg One time	0		0	0	0	0
10-45 days	35 days	0	8.2 kg	0.51 kg/2 days (16 doses)	33.0 kg	1.94 kg/2 days (17 doses)	0	0
46-90 days	45 days	0	8.2 kg	0.51 kg/2 days (16 doses) till 80 DAS	33.0 kg	1.5 kg/2 days (22 doses)	0	0
91-140 days	50 days	0	0	0	16.5 kg	1.1 kg/2 days (15 doses) till 120 DAS	33.4 kg	1.3 kg/2 days (25 doses)

2. Black Pepper (*Piper nigrum*)

RD# 32:28:110 kg/acre NPK.
(from Year 3 onwards)

#The recommendation is based on the target yield and *in situ*soil fertility (IISR). I have prepared the schedule for a medium fertility soil for a target yield of 2.5 t/acre dry pepper. The soil fertility levels are as follows:

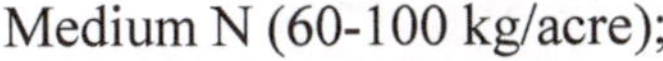

Medium N (60-100 kg/acre);

Medium P (4-12 kg/acre); and

Medium K (44-120 kg/acre).

Ratio

Crop Stages	Duration	N	P	K
May-June	60 days	0.6	0.7	0.2
Aug-Sept	60 days	0.4	0.3	0.9

Fertigation schedule

Crop Stages	Duration	Urea kg/acre	Urea schedule	Phosphoric Acid kg/acre	Phosphoric Acid Schedule	MOP kg/ acre	MOP Schedule
May-June	60 days	42	2.1 kg/3 days/acre (20 doses)	37.6	1.9 kg/3 days/acre (20 doses)	36.7	1.8 kg/3 days/acre (20 doses)
Aug -Sept.	60 days	28	1.4 kg/3 days/acre (20 doses)	16.1	0.8 kg/3 days/acre (20 doses)	147	7.4 kg/3 days/acre (20 doses)

Apply 1 kg/plant lime (powdered) to the base of each vine for soil pH correction before start of fertigation or after the first period of fertigation during the break time.

Apply 2 kg powdered dry cow dung or compost +1 kg neem cake to each vine at the base (x 4 doses /year).

Apply 5 g/vine of MgSO4 in the first week and last week of June. Apply to the soil or add to the fertigation solution (the 10th and 20th doses during May-June period).

Apply 5g/vine of Calcium Nitrate in the first week and last week of September. Apply to the soil.

Micro nutrient spray (one spray every month).

Boron 5 ppm; Ferrous Sulphate 15 ppm; Manganese sulphate 10 ppm; Zinc Sulphate 15 ppm; and Copper sulphate 5 ppm.

3. Chilly (Red) (*Capsicum frutescens*)

RD 145:82:185 kg/acre NPK (Prakasam dist. AP)

Ratio

Growth stage	Duration	N	P	K
Basal	At Land preparation	0	0.75	0
10 -30 DAP	20 days	0.15	0	0
31-60 DAP	30 days	0.35	0.15	0.1
61-90 DAP	30 days	0.4	0.1	0.45
91-120 DAP	30 days	0.1	0	0.45

Fertilizers

Single superphosphate (SSP)	384 kg/acre (Apply direct to soil)
Magnesium sulphate (MgSO4)	10 kg/acre (Apply direct to soil)
Mono Ammonium Phosphate (MAP)	33.6 kg/acre
Urea	274.6 kg/acre
Ammonium Sulphate	69 kg/acre
White Potash (MOP)	169.9 kg/acre (101.9 K)
Potassium sulphate (SOP)	166.5 kg/acre (83.3 K)

Fertigation schedule

Growth Stages	Duration	Fertilizer	Total Quantity kg/acre	Dose
Basal	At Bed preparation time	SSP	384	Full at final Bed preparation (+ Mg SO4, Organic Manure and PSB)
10-30 DAP	20 days	Urea	47.2	2.36 kg/day/acre (20 doses)
31-60 DAP	30 days	MAP	20.2	1.35 kg/2 days/acre (15 doses)
		Urea	105	7 kg/2 days/acre (15 doses)
		White MOP	30.9	2.1 kg/2 days/acre (15 doses)
61-90	30 days	MAP	13.45	0.9 kg/2 days/acre (15 doses)
		Urea	122.4	8.16 kg/2 days/acre (15 doses)
		MOP	139	9.3 kg/2 days/acre 15 doses)
91-120	30 days	Ammonium Sulphate	69	4.6 kg/2 days/acre (15 doses)
		SOP	166.5	11.1 kg/2 days/acre (15 doses)

MgSO4 10 kg/acre as Basal along with SSP.

4. Ginger (*Zingiberofficinale*)

RD 45:32:48 kg/acre NPK (Sangareddy Res station, Telengana)

Ratio

Growth stages	Duration	N	P	K
Basal	At Land preparation	0	0.7	
15-30 DAE	15 days	0.24		
31-90 DAE	60 days	0.24	0.3	0.1
91-120 DAE	30 days	0.27		0.25
121-170 DAE	50 days	0.21		0.35
171-220 DAE	50 days	0		0.3

Fertilizers

- SSP (Apply direct to soil)

- MAP
- Urea
- Calcium Nitrate
- MOP White
- SOP
- Magnesium Sulphate

Fertigation schedule

Growth stages	Duration	Fertilizer	Total Quantity kg/acre	Schedule
Basal	At Ridge preparation	SSP	140	
15-30 DAE	15 days	Urea	24	1.6 kg/day/acre (15 doses)
31-90 DAE	60 days	MAP	15.74	0.52 kg/2 days/acre (30 doses)
		Urea	23	0.77 kg/2 days/acre (30 doses)
		MOP white (from 70 DAE only)	8.8	0.9 kg/2 days /acre (10 doses)
91-120 DAE	30 days	Urea	26	1.7 kg/2 days/acre (15 doses)
		MOP white	20	1.3 kg/2 days/acre (15 doses)
		MgSO4	25	1.7 kg/2 days/acre (15 doses)
121-170 DAE	50 days	Urea (till 140 DAE only)	12.5	1.25 kg/2 days/acre (10 doses)
		Calcium Nitrate	25	1 kg/2 days/acre (25 doses)
		MOP white	27.5	1.1 kg/2 days/acre (25 doses)
171-220 DAE	50 days	SOP (till 210 DAE only)	28	1.4 kg/2 days/acre (20 doses)

 # Micronutrients: 2 sprays of Zn 0.3%, Fe 0.2% and B 0.2% during 45-60 DAE and repeat 2 sprays 75-90 DAE

5. Turmeric (*Curcuma longa*)

RD 60:50:100 kg NPK/acre

Ratio

Growth Stages	Duration	N	P	K
At land preparation	Before Planting	0	0.4	0
10 -80 DAE	70 days	0.2	0	0
81-160 DAE	80 days	0.4	0.3	0.3
161-200 DAE	40 days	0.2	0.3	0.5
201-220 DAE	20 days	0	0	0.2

Fertilizers

- SSP (Apply direct to soil)
- MAP
- Urea
- MOP white

Fertigation schedule

Growth stages	Duration	Fertilizer	Total Quantity kg/acre	Schedule
Basal	At Bed preparation	SSP	125	One time direct to soil
10-80 DAE	70 days	Urea	26	0.75 kg/2 days/acre (35 doses)
81-160 DAE	80 days	MAP (from 40 DAE only)	24.6	1.2kg/2 days/acre (20 doses)
		Urea	45.6	1.1 kg/2 days/acre (40 doses)
		White MOP (from 100 DAE only)	50	1.7 kg/2 days /acre (30 doses)
161-200 DAE	40 days	MAP	24.6	1.2 kg/2 days/acre (20 doses)
		Urea (till 180 DAE only)	19.5	1.9 kg/2 days/acre (10 doses)
		White MOP	83.5	4.2 kg/2 days/acre (20 doses)
201-220 DAE	20 days	White MOP	33.4	3.3 kg/2 days/acre (10 doses)

Apply $MgSO_4$ and micro nutrients as per requirement (soil test)

6. Curry leaf (*Murraya koenigii*)

RD 100:100:100 kg/acre NPK

Ratio

Growth stages	Duration	N	P	K
Basal at planting	Planting time	0.08	0.2	0
30-345 days	First year growth	0.12	0	0.2
First harvest (after 1 yr of growth)	One harvest at every 45 days & fertigation at the end of each harvest in the proceeding 45 days period.	0.1	0.1	0.1
Repeat after each cut (total 8 harvests every year)		0.1x 8	0.1 x 8	0.1 x 8

Fertilizers

- DAP (Apply direct to soil)
- Urea
- MOP
- 19:19:19

Fertigation schedule

Stages of growth	Duration	Fertilizer	Quantity kg/acre	Schedule
Basal	At Planting	DAP	43.5	soil application
30-345 days (yr1)	315 days	Urea	26.5	0.6 kg/week /acre (45 doses)
		White MOP	33.4	0.74 kg/week/acre (45 doses)
yr2 1-45 Days after first harvest preparing for harvest 2	45 days	19:19:19	52.6	2.4 kg/2 days/acre (22 doses)
Every 45 days a harvest and fertigation during 45 days before each harvest	For example, 46-91 days fertigation	19:19:19	52.6	2.4 kg/2 days/acre (22 doses)

$$Repeat the same schedule after 3, 4, 5, 6, 7, 8 harvests in a year (and there is 8 such harvests/ year. ##8 times 10% (0.1 NPK) each for 8 harvests in a year.

7. Cardamom (*Elettaria cardamomum*)

RD 30:30:60 kg/acre NPK (yr 2-4)

60:60:120 kg/acre NPK (yr 5-10 plus)

Ratios

Year 2-4

Fertigation period	days	N	P	K
April-May-June-July-August	150	0.5	0.7	0.4
Sept 15th -October-Nov-Dec-Jan-Feb	165	0.5	0.3	0.6

Year 5-10 plus

Fertigation period	days	N	P	K
April-May-June-July-August	150	0.5	0.7	0.4
Sept 15th -October-Nov-Dec-Jan-Feb	165	0.5	0.3	0.6

Fertigation schedule

Fertigation period (fertigate once in two days)	Year 2 - 4 (Age)								
	No of doses/ year	Urea kg/ dose	Total Urea kg/ha	19:19 :19 kg/ dose	Total 19:19 :19 kg/ha	MAP kg/ dose	Total MAP kg/ha	White MOP kg/ dose	Total MOP kg/ha
April to August (150 days)	75 /year	0.33	24.75	1.33	100	0.73	55	0.91	68.3
Sept 15 to Feb (165 days)	83/year	0.39	32.6	1.43	118.5	0	0	1.36	112.7
	Year 5 to Year 10 plus								
April to August	75/year	0.69	51.6	2.67	200	1.47	110	1.83	137
Sept 15 to Feb	83/year	0.78	65.1	2.86	237	0	0	2.72	225.5

Micronutrients - Zn (250 g/100 l water rate ZnSO4. Spray once in both periods above.

Boron - Soil application 7.5 kg/ha Borax ;50% period 1 and 50% period 2.

IISR micro mixture 5 g/liter water two times (1.period 1; 2.period 2)

G. INDUSTRIAL CROPS

1. SUGAR CANE (*Saccharumofficinarum*)

RD 100:30:75 kg/acre NPK (north Karnataka, University Agriculture, Dharwad)

Ratio

Sugar cane Growth Stages	No of days	N	P	K
16 -45 DAE	30	0.15	0.15	0
46-90 DAE	45	0.35	0.5	0.1
91-180 DAE	90	0.5	0.35	0.4
181-265 DAE	85	0		0.5

Fertilizers

- Urea
- MAP
- White MOP

Fertigation Schedule

Growth stage	Duration	Fertilizer	Total Quantity kg/acre	Schedule
16-45 DAE	30 days	Urea	32.6	2.2 kg/2 days/acre (15 doses)
		MAP	7.4	0.5 kg/2 days/acre (15 doses)
46-90 DAE	45 days	Urea	76	3.5 kg/2 days/acre (22 doses)
		MAP	24.6	1.1 kg/2 days/acre (22 doses)
		White MOP	12.5	0.6 kg/2 days/acre (22 doses)
91-180 DAE	90 days	Urea	108.5	2.4 kg/2 days/acre (45 doses)
		MAP	17.2	0.4 kg/2 days/acre (45 days)
		White MOP	50.1	1.1 kg/2 days/acre (45 doses)
181-265 DAE	85 days	White MOP	62.6	1.5 kg/2 days/acre (42 doses)

Secondary and micro nutrients

	Kg/acre
FeSO4	10
ZnSO4	10
MnSO4	5
Borax	2
MgSO4	50

#50% of all micronutrients and MgSO4 apply as basal dose along with the FYM (Organic manure at land preparation)
#50% of all micronutrients and MgSO4 apply direct to soil at Earthing up (75-90 DAE)

2. Cotton (*Gossypium* spp.)

RD 50:30:35 NPK kg/acre (Tamilnad)

Ratio

Growth Stage	Duration	N	P	K
Basal			0.7	
5-30 DAE	25 days	0.3	0	0.1
31-60 DAE	30 days	0.6	0.3	0.5
61-100 DAE	40 days	0.1		0.4

Addition of Mg SO_4 and micro nutrients

Fertigation schedule

Growth stage	Duration	Fertilizer	Total Quantity kg/acre	Schedule
At Land preparation	Basal	SSP	131.3	Soil application
5-30 DAE	25 days	Urea	32.6	2.7 kg/2 days/acre (12 doses)
31-60 DAE	30 days	Urea	65	4.3 kg/2 days/acre (15 doses)
		MAP	15	1 kg/2 days/acre (15 doses)
		White MOP	29.2	1.9/2 days/acre (15 doses)
61-100 DAE	40 days	Urea till 80 DAE only	7.4	0.74 kg/2 days/acre (10 doses)
		White MOP	29.3	1.5 kg/2 days/acre (20 doses)

Mg SO_4 10 kg/acre to be applied during 31-60 DAE

3. Tobacco (*Nicotiana tabaccum*)

RD 36:14:48 NPK and 6 (Ca) kg/acre (Northern Light soils West Godavari, AP) CTRI -ICAR

Ratio

Growth stages	Duration	N	P	K
Basal dose	At Land preparation	0.15	1	0.1
Week 2 (after germination)	7 days	0.1	0	0.15
Week 3	7	0.2	0	0.15
Week 4	7	0.2	0	0.15
Week 5	7	0.2	0	0.15
Week 6	7	0.1	0	0.15
Week 7	7	0.05	0	0.15

##Fertigation should be completed by 40-45 days of growth

Fertilizers

- DAP
- Urea
- Ammonium sulphate
- Potassium Sulphate

Fertigation schedule

Growth stages	Duration	Fertilizer	Total quantity kg/acre	Schedule
Basal dose	At Land preparation	DAP	30.4	Apply to Soil at planting
Week 2 (after germination)	7 days	Urea	7.8	2.6 kg/ 2 days /acre (3 doses)
		SOP	14.4	4.8 kg/2 days/acre (3 doses)
Week 3	7	Urea	15.6	5.2 kg/2 days/acre (3 doses)
		SOP	14.4	4.8 kg/2 days/acre (3 doses)
Week 4	7	Urea	15.6	5.2 kg/2 days/acre (3 doses)
		SOP	14.4	4.8 kg/2 days/acre (3 doses)
Week 5	7	Urea	3	3 kg/2 days/acre (1 dose)
		Ammo.Sulphate	27.6	9.2 kg/2 days/acre (3 doses)
		SOP	14.4	4.8 kg/2 days/acre (3 doses)
Week 6	7	Ammo. Sulphate	17.1	5.7 kg/2 days/acre (3 doses)
		SOP	14.4	4.8 kg/2 days/acre (3 doses)
Week 7$$	7	Ammo. Sulphate	8.6	4.3 kg/2 days/acre (2 doses)
		SOP	14.4	7.2 kg/2 days/acre (2 doses)

$$Last fertigation on day 46.

H. MEDICINAL CROPS

1. Stevia (*Stevia rebaudiana*)

R 12:44:44
NPK kg/acre (Punjab)

Ratio

Growth stages	Duration	N	P	K
10 -90 DAE	80 days	0.2	0.3	0.2
91- 120 DAE	30 days	0.2	0.25	0.2
150- 180 DAE	30 days	0.2	0.15	0.2
181- 220 DAE	40 days	0.2	0.15	0.2
240-320 DAE	80 days	0.2	0.15	0.2

Basal dose at planting Organic manure 8 t/acre

Fertilizers

- Urea
- Phosphoric Acid
- MOP (white)
- And $MgSO_4$.

Fertigation schedule

Growth stages	Duration	Fertilizer	Total Quantity kg/acre	Schedule
10 -90 DAE	80 days	Urea	5.2	0.5 kg/week/acre (11 days)
		Phosphoric acid	25.3	2.3 kg/week/acre (11 doses)
		MOP White	14.7	1.3 kg/week/acre (11 doses)
91- 120 DAE	30 days	Urea	5.2	1.3 kg/week/acre (4 doses)
		Phosphoric acid	21.1	5.3 kg/week/acre (4 doses)
		MOP White	14.7	3.7 kg/week/acre (4 doses)
150- 180 DAE	30 days	Urea	5.2	1.3 kg/week/acre (4 doses)
		Phosphoric acid	12.7	3.2 kg/week/acre (4 doses)
		MOP White	14.7	3.7 kg/week/acre (4 doses)
181- 220 DAE	40 days	Urea	5.2	1 kg/week/acre (5 doses)
		Phosphoric acid	12.7	2.5 kg/week/acre (5 doses)
		MOP White	14.7	2.9 kg/week/acre (5 doses)
240-320 DAE	80 days	Urea	5.2	0.5 kg/week/acre (11 doses)
		Phosphoric acid	12.7	1.2 kg/week/acre (11 doses)
		MOP White	14.7	1.3 kg/week/acre (11 doses)

I. FIELD GROWN FLOWERS

1. Chrysanthemum (*Chrysanthemum morifolium*)

RD **50:48:10 NPK kg/acre**

Ratio

Growth stages	Duration	N	P	K
At Planting	Basal	0.25	0.6	0
21-40 DAE	20 days	0.4	0.3	0.6
41-65 DAE	25 days	0.1	0	0.3

Azospirillum and Phospho- bacterium; 2 kg each to soil at planting

Foliar spray of ZnSO4 (0.25%) and MgSO4 (0.25 %)

Fertilizers

- SSP
- Urea
- Phosphoric acid
- SOP

Fertigation schedule

Growth stages	Duration	Fertilizer	Total quantity kg/acre	Schedule
At Planting	Basal	SSP	180	At planting
		Urea	27	At planting
10-20 DAE	10 days	Urea	27	5.4 kg/2 days/acre (5 doses)
		Phosphoric acid	9.2	1.8 kg/2 days/acre (5 doses)
		SOP	2	0.4 kg/2 days/acre (5 doses)
21-40 DAE	20 days	Urea	43.4	4.3 kg/2 days/acre (10 doses)
		Phosphoric acid	27.6	2.8 kg/2 days/acre (10 doses)
		SOP	12	1.2 kg/2 days/acre (10 doses)
41-65 DAE	25 days	Urea	11	0.9 kg/2 days/acre (12 doses)
		SOP	6	0.5 kg/2 days/acre (12 doses)

2. Jasmine (*Jasminum* spp.)

RD 126:312:312
NPK kg/acre/year

Ratio

Growth stage /Time	Duration	N	P	K
June 1st week	Soil apply		0.5	
June-August	92 days	0.6		0.5
November (after pruning)	Soil apply		0.5	
December-February	90 days	0.4		0.5

Fertilizers

- SSP
- Urea
- Ammonium sulphate
- Potassium nitrate

Fertigation schedule (for standing crop)

Growth stages	Duration	Fertilizer	Total quantity kg/acre	Schedule
June 1st week	Soil application	SSP	156 kg/acre (375 g/plant)	Direct to soil
June-August	92 days	Urea (first 40 days)	54.3 kg/acre	2.7 kg/2 days/acre (first 20 doses)
		Ammonium Sulphate (41-92 days)	119 kg/acre	4.6 kg/2 days/acre (26 doses)
		Pot. Nitrate (full duration)	338.5 kg/acre	7.4 kg/2 days/acre (46 doses)
November (after pruning)	Soil application	SSP	156 kg/acre (375 g/plant)	Direct to soil
December-February	90 days	Urea (first 40 days)	19.5 kg/acre	1kg/2 days/acre (20 doses)
		Ammonium sulphate (41-90 days)	43 kg/acre	1.7 kg/2 days/acre (25 doses)
		Pot.Nitrate (full duration)	338.5 kg/acre	7.5 kg/2 days/acre (45 doses)

Schedule is repeated every year.

$$Micro nutrient foliar spray- 0.25 % $ZnSO_4$, 0.5 % $MgSO_4$, 0.5% $FeSO_4$. One spray every 15 days (June-August and December-February)

J. PLANTATION CROPS

1. Tea (*Camellia sinensis*)

RD Recommendation by UPASI (United Planters Association of South India)

Made tea (kg/acre)	N (kg/acre) O.M. status		
	Low	Medium	High
Minimum	64	48	40
800	96	80	72
1000	106	90	80
1200	116	100	88
1400	126	110	96
1600	136	120	104

Nitrogen is based on target yield

It also varies with the OM (organic matter status of the soil)

Fertigation Ratio and Schedulefor Nitrogen

The schedule given here is for Medium OM soil and 1200kg/acre yield target (N= 100 kg/acre)&&

Growth time Pre-monsoon	Duration	N (fraction)	Fertilizer	Fertigation schedule
(April-May)	60 days	0.2	Ammonium Sulphate for 20 N = 95.2 kg/acre	3.1 kg/2 days/acre (30 doses)
Jun-July	60	0.2	Urea for 20N = 43.4 kg/acre	1.4 kg/2 days /acre (30 doses)
Aug-sept	60	0.3	Urea for 30 N = 65.1 kg/acre	2.2 kg/2 days/acre (30 doses)
Oct-Nov	60	0.2	Urea for 20N = 43.4 kg/acre	1.4 kg/2 days/acre (30 doses)
Dec-Jan 15	45	0.1	CAN for 10 N = 38.5 kg/acre	1.8 kg/2 days/acre (22 doses)

&& It is recommended to use Ammonium sulphate source of N in Pre monsoon.
Then Urea during the monsoon period and CAN during winter months
P is applied to the soil as Rock phosphate.
Ca is applied through periodical liming of the soil.
K fertilizer is applied based on N:K ratio followed after pruning and varies with plantations
Mg as $MgSO_4$

2. Coffee (*Coffea* spp.)

RD 120:90:120 NPK kg/acrefor Arabica (Central Coffee Research Institute-Karnataka)

Ratio

Growth stage	Duration	N	P	K
January	30 days	0	0	0
##Feb 15th -April	74 days	0.25	0.4	0.15
May	31 days	0.15	0.3	0.15
June 1-30	30days	0.2	0	0.15
July	31 days	0	0	0
16-31 August	15 days	0.15	0	0.15
Sept-December	122 days	0.25	0.3	0.4

Fertilizers

- Urea
- MAP
- MOP

Fertigation schedule

Growth stage	Duration	Urea kg/acre	Schedule	MAP kg/acre	Schedule	MOP kg/acre	Schedule
January	30 days	0		0		0	
##Feb 15th -April	74 days	49.7	2 kg/3 days/acre (25 doses)	59	2.4 kg/3 days/acre (25 doses)	30.1	1.2 kg/3 days/acre (25 doses)
May	31 days	27.6	2.8 kg/3 days/acre (10 doses)	44.3	4.4 kg/3 days/acre (10 doses)	30.1	3.0 kg/3 days/acre (10 doses)
June 1-30	30 days	52.1	5.2 kg/3 days/acre (10 doses)	0		30.1	3.0 kg/3 days/acre (10 doses)
July	31 days	0	0	0		0	
16-31 August	15 days	39.1	7.8 kg/3 days/acre (5 doses)	0		30.1	6.0 kg/3 days/acre (5 doses)
Sept-December	122 days	53.6	1.3 kg/3 days/acre (40 doses)	44.3	1.1 kg/3 days/acre (40 doses)	80.2	2.0 kg/3 days/acre (40 doses)

Pruning of coffee bushes done in first week of February

K. FRUIT CROPS (Annual & Seasonal)

1. Banana (*Musa paradisiaca*)

RD 220:70:400 NPK kg/acre (Tamilnadu)

Ratio

Growth Stage	Duration	N	P	K
7-66 DAP	60 days	0.15	0.3	0
67-127 DAP	60 days	0.25	0.4	0.1
128-187 DAP	60 days	0.25	0.3	0.2
188-307 DAP	120 days	0.35		0.7

Fertilizers

- Urea
- MAP
- MOP

Fertigation schedule

Growth Stage	Duration	Urea	Schedule	MAP	Schedule	MOP	Schedule
7-66 DAP#	60 days	62.7	2.1 kg/2 days/acre (30 doses)	34.4	1.1 kg/2 days/acre (30 doses)	0	
67-127 DAP	60 days	107.4	3.6 kg/2 days/acre (30 doses)	45.9	1.5 kg/2 days/acre (30 doses)	66.8	2.2 kg/2 days/acre (30 doses)
128-187 DAP	60 days	110.5	3.7 kg/2 days/acre (30 doses)	34.4	1.1 kg/2 days/acre (30 doses)	133.6	4.5 kg/2 days/acre (30 doses)
188-307 DAP	120 days	167.9	5.6 kg/2 days/acre (30 doses) till 248 DAP	0		467.6	7.8 kg/2 days/acre (60 doses)

DAP = Days after Planting

Secondary Nutrients

100 kg MgSO4 /acre – to be fertigated from 67 DAP to 187 DAP (120 days) every 2 days (60 doses) at 1.7 kg/acre/dose.

10 kg Ca (NO3)2/acre - to be fertigated 1 kg/acre per dose every 14 days from 67 DAP.

Micro nutrients

20 kg FeSO4 and 20 kg ZnSO4 to be fertigated 1 kg each every 14 days from 67 DAP (Use chelate of iron instead of FeSO4 for fertigation).

2. Plantain (*Nendran*) (*Musa* spp.)

RD 250:150:500 NPK kg/acre

Ratio

Growth stages	Duration	N	P	K
7-60 DAP	53 days	0.2	0.2	0
61-120 DAP	60 days	0.3	0.4	0.05
121-180 DAP	60 days	0.3	0.3	0.05
181-270 DAP	90 days	0.2	0	0.1
271-330 DAP	60 days	0	0.1	0.3
331-390 DAP	60 days	0	0	0.4
391-420 DAP	30 days	0	0	0.1

Fertilizers

- Urea
- MAP
- MOP

Fertigation schedule

Growth stage	Duration	Urea kg/acre	Schedule	MAP kg/acre	Schedule	MOP kg/acre	Schedule
7-60 DAP	53 days	95.7	3.7 kg/2 days/acre (26 doses)	49.2	1.9 kg/2 days/acre (26 doses)	0	
61-120 DAP	60 days	135	4.5 kg/2 days/acre (30 doses)	98.4	3.3 kg/2 days/acre (30 doses)	41.8	1.4 kg/2 days/acre (30 doses)
121-180 DAP	60 days	143.4	4.8 kg/2 days/acre (30 doses)	73.8	2.5 kg/2 days/acre (30 doses)	41.8	1.4 kg/2 days/acre (30 doses)
181-270 DAP	90 days	108.5	2.4 kg/2 days/acre (45 doses)	0		83.6	1.9 kg/2 days/acre (45 doses)
271-330 DAP	60 days	0		24.6	0.8 kg/2 days/acre (30 doses)	250.5	8.4 kg/2 days/acre (30 doses)
331-390 DAP	60 days	0		0		334	11.1 kg/2 days/acre (30 doses)
391-420 DAP	30 days	0		0		83.6	5.6 kg/2 days/acre (15 doses)

DAP = Days after Planting

Secondary Nutrients

100 kg $MgSO_4$ /acre – to be fertigated from 60 DAP to 180 DAP (120 days) every 2 days (60 doses) at 1.7 kg/acre/dose.

10 kg $Ca(NO_3)_2$/acre - to be fertigated 1 kg/acre per dose every 14 days from 121 DAP.

Micro nutrients

20 kg $FeSO_4$ and 20 kg $ZnSO_4$ to be fertigated 1 kg each every 14 days from 60 DAP (Use chelate of iron instead of $FeSO_4$ for fertigation).

3. Watermelon (*Citrullu slanatus*)

RD 80:50:60 NPK kg/acre

Ratio

Growth stage	Duration	N	P	K
Basal	Soil application	0.2	0.9	0
10-40 DAE	30 days	0.4	0	0.15
41- 80 days	40 days	0.4	0.1	0.65
81-90 days	10 days	0		0.2

Fertigation schedule

Growth stage	Duration	DAP (kg/acre)	DAP application	Urea (kg/acre)	Schedule of Urea	Phosphoric Acid (kg/acre)	Schedule of Phosphoric acid	MOP (kg/acre)	Schedule of MOP
Basal	Soil application	100	At Land preparation						
10-40 DAE	30 days			67.3	6.7 kg/3 days/acre (10 doses)	0		15	1.5 kg/3 days /acre (10 doses)
41- 80 days	40 days			67.3	5.2 kg/3 days/acre (13 doses)	7.7	0.6 kg/3 days/acre (13 doses)	65	5 kg/3 days/acre (13 doses)
81-90 days	10 days			0		0		20	2 kg/3 days/acre (10 doses)

L. FRUIT CROPS (Perennial)

1. Mango (*Mangi feraindica*)

Mango Fertigation Schedule for 6 year Plus Trees at 10 x 10 m crop geometry

RD 1:1:1 NPK kg/tree/yr

Ratio

Growth Stages	Duration	N	P	K
June 4th week-July 3rd week	30 days	0.5	0.5	0.5
September	##one soil application	0	0.25	0.5
October	30 days	0.5	0	0
November	one soil application	0	0.25	0

##Soil application of SSP

*All 6.25 kg SSP to be applied in three splits, June (3.1 kg /tree). Sept. (1.6kg/tree), November (1.6 kg/tree) as soil application.

**Apply the SSP in two rings around the tree 1m and 2m away from the trunk, respectively.

**Place the fertilizer at a depth of 10-15 cm below the soil surface and cover with top soil.

Fertigation schedule

Growth Stages	Duration	Urea g/tree	Schedule	SSP g/tree (Soil application)##	MOP g/tree	Schedule
June 4th week-July 3rd week	30 days	1050	105 g/3 days/tree (10 doses)	3125	835	83.5 g/3 days/ tree (10 doses)
September		0		1600	0	
October	30 days	1050	105 g/3 days/tree (10 doses)	0	835	83.5 g/3 days/ tree (10 doses)
November		0		1600	0	

1. a) Mango (Ultra High density planted at 3 x 2 m)

RD (This recommendation is based on Our experience in Jain R& Dfarm Tamilnad).

year 1 23:10:12 NPK kg/acre

year 2 30:15:33 kg/acre

year 3 57:29:51 kg/acre

year 4 plus 72:36:66 kg/acre

Ratio

Growth stage	Duration	N	P	K
yr1	July-Sept	0.3	0.3	0.5
	Jan-May	0.7	0.7	0.5

Fertigation schedule

Growth Stages	Duration	Urea kg/acre	Schedule	H3PO4 kg/acre	Schedule	MOP kg/acre	Schedule
July-September	90 days	15	0.5 kg/3 days/acre (30 doses)	5.8	0.2 kg/3 days/acre (30 doses)	10	0.3 kg/3 days/acre (30 doses)
Jan-May	150 days	35	0.7 kg/3 days/acre (50 doses)	13.4	0.27 kg/3 days/acre (50 doses)	10	0.5 kg/3 days/acre (20 doses in March-April)

Ratio

Growth stage	Duration	N	P	K
yr2	July-Sept	0.5	0.5	0.5
	Jan-May	0.5	0.5	0.5

Fertigation schedule

Growth Stages	Duration	Urea kg/acre	Schedule	H3PO4 kg/acre	Schedule	MOP kg/acre	Schedule
July-September	90 days	32.6	1.1 kg/3 days/acre (30 doses)	14.4	0.48 kg/3 days/acre (30 doses)	27.6	0.9 kg/3 days/acre (30 doses)
Jan-May	150 days	32.6	0.65 kg/3 days/acre (50 doses)	14.4	0.3 kg/3 days/acre (50 doses)	27.6	0.6 kg/3 days/acre (50 doses)

Ratio

Growth stage	Duration	N	P	K
yr3	June15-Aug	0.45	0.5	0.5
	Sept	0.05	0.1	0.15
	Jan-May	0.5	0.4	0.35

Fertigation schedule

Growth Stages	Duration	Urea kg/acre	Schedule	H_3PO_4 kg/acre	Schedule	MOP kg/acre	Schedule
June15-August	77 days	52.7	2.1 kg/3 days/acre (25 doses)	27.8	1.1 kg/3 days/acre (25 doses)	42.6	1.7 kg/3 days/acre (25 doses)
September	30 days	6.2	0.6 kg/3 days/acre (10 doses)	5.6	0.6 kg/3 days/acre (10 doses)	12.8	1.3 kg/3 days/acre (10 doses)
Jan-May	150 days	61.8	1.2 kg/3 days/acre (50 doses)	22.3	0.45 kg/3 days/acre (50 doses)	29.8	0.6 kg/3 days/acre (50 doses)

Ratio

Growth stage	Duration	N	P	K
yr 4 onwards	June15-Aug	0.55	0.6	0.5
	Sept	0.05	0.1	0.15
	Jan-May	0.4	0.3	0.35

Fertigation schedule

Growth Stages	Duration	Urea kg/acre	Schedule	H_3PO_4 kg/acre	Schedule	MOP kg/acre	Schedule
June15-August	77 days	85.9	3.4 kg/3 days/acre (25 doses)	41.5	1.4 kg/3 days/acre (25 doses)	55.1	1.8 kg/3 days/acre (25 doses)
September	30 days	7.8	0.8 kg/3 days/acre (10 doses)	6.9	0.7 kg/3 days/acre (10 doses)	16.5	1.7 kg/3 days/acre (10 doses)
Jan-May	150 days	62.5	1.3 kg/3 days/acre (50 doses)	20.7	1 kg/3 days/ acre (20 doses-during January-February)	38.6	0.8 kg/3 days/acre (50 doses)

Micro nutrients-Apply standard micro nutrient mixture (available in the market). 4 x 50g/acre doses in July and 4 x 50 g/acre doses in November as foliar spray in from year 3 onwards.

Apply MgSO4 thru fertigation 10 kg/acre in 20 doses (yr 2, and yr 3) June-August period And 20 kg/acre in 30 doses (yr 4 onwards) August- November period.

2. Cashew (*Annacardium occidentale*)

RD Recommended by TNAU

Year	N kg/acre	P kg/acre	K kg/acre
1	18	8	6
2	36	16	12
3	54	24	18
4	72	33	25
5	90	41	30

Ratio

Year of growth	Growth Stage	Duration	N	P	K
1	July-Sept	92 days	0.3	0.4	0.4
	Jan-May	160 days	0.7	0.6	0.6
2	July-Sept	92 days	0.35	0.35	0.35
	Jan-May	160 days	0.65	0.65	0.65
3	July-Sept	92 days	0.35	0.35	0.35
	Jan-May	160 days	0.65	0.65	0.65
4	Jun-Aug	92 days	0.5	0.5	0.5
	Sept	30 days	0.15	0.2	0.2
	Jan-March	90 days	0.35	0.3	0.3
5	Jun-Aug	92 days	0.5	0.5	0.5
	Sept	30 days	0.15	0.25	0.15
	Jan-March	90 days	0.35	0.25	0.35

Fertigation schedule

Year of growth	Growth Stage	Duration	Urea kg/ac	Schedule	Phosphoric acid	Schedule	MOP	Schedule
1	July-Sept	92 days	13.2	0.6 kg/4 days/acre (23 doses)	6	0.26 kg/4 days/acre (23 doses)	3.6	0.3 kg/4 days/acre (Aug-Sept only 12 doses)
	Jan-May	160 days	26	0.65 kg/4 days/acre (40 doses)	10	0.25 kg/4 days/acre (40 doses)	6	0.3 kg/4 days/acre (Mar-May only 20 doses)
2	July-Sept	92 days	27.6	1.2 kg/4 days/acre (23 doses)	10.8	0.47 kg/4 days/acre (23 doses)	7.2	0.6 kg/4 days/acre (Aug-Sept only 12 doses)
	Jan-May	160 days	50	1.25 kg/4 days/acre (40 doses)	20	0.5 kg/4 days/acre (40 doses)	12	0.6 kg/4 days/acre (Mar -May only 20 doses)
3	July-Sept	92 days	40.8	1.8 kg/4days/acre (23 doses)	16.8	0.7 kg/4 days/acre (23 doses)	10.8	0.9 kg/4 days/acre (Aug-Sept only 12 doses)
	Jan-May	160 days	76	1.9 kg/4 days/acre (40 doses)	30	0.75 kg/4 days/acre (40 doses)	20	1 kg/4 days/acre (Mar-May only 20 doses)
4	Jun-Aug	92 days	78	3.4 kg/4 days/acre (23 doses)	31.2	1.4 kg/4 days/acre (23 doses)	20.4	1.7 kg/4 days/acre (Jul-Aug only 12 doses)
	Sept	30 days	23.6	3.4 kg/4 days/acre (7 doses)	15.6	2.2 kg/4 days/acre (7 doses)	6	0.86 kg/4 days/acre (7 doses)
	Jan-March	90 days	55.2	2.5 kg/4 days/acre (22 doses)	16.9	0.77 kg/4 days/acre (22 doses)	14.4	1.4 kg/4 days/acre (Feb-Mar only (10 doses)
5 onwards	Jun-Aug	92 days	97.2	4.2 kg/4 days/acre (23 doses)	39.6	1.7 kg/4 days/acre (23 doses)	25.2	2.1 kg/4 days/acre (Jul-Aug only 12 doses)
	Sept	30 days	29.2	4.2 kg/4 days/acre (7 doses)	19.6	2.8 kg/4 days/acre (7 doses)	7.6	1.1 kg/4 days/acre (7 doses)
	Jan-March	90 days	68.4	3.1 kg/4 days/acre (22 doses)	19.2	0.87kg/4 days/acre (22 doses)	18	1.8 kg/4 days/acre (feb-Mar only 10 doses)

##Micronutrients, Standard micro nutrient mixture 1 kg /time x 2 to be fertigated

3. Aonla or Indian Gooseberry (*Phyllanthu semblica*)

Pre-bearing stage (year 2 & 3)

RD 50:40:50 kg/acre NPK

Ratio

Growth stage	Duration	N	P	K
Sep-October	60 days	0.5	0.5	0.5
April -May	60 days	0.5	0.5	0.5

Fertigation Schedule

Growth stage	Duration	Urea kg/ acre	Schedule	SSP kg/ acre	Schedule	White MOP kg/acre	Schedule
Sep-October	60 days	54.3	2.7 kg/3 days/acre (20 doses)	125	Soil application in two rings (0.65 kg/tree)	41.8	2.1 kg/3 days/acre (20 doses)
April-May	60 days	54.3	2.7 kg/3 days/acre (20 doses)	125	Soil application in two rings (0.65 kg/tree)	41.8	2.1 kg/3 days/acre (20 doses)

Bearing stage (Year 4 onwards)

RD 145:100:190 NPK kg/acre

Ratio

Growth stage	Duration	N	P	K
Aug-Oct 15	75 days	0.4	0.5	0.3
March	30 days	0.2		0.2
April-June 15	75 days	0.4	0.5	0.5

Fertigation Schedule

Growth stage	Duration	Urea kg/ acre	Schedule	SSP kg/ acre	Schedule	White MOP kg/acre	Schedule
Sep-October	60 days	126	6.3 kg/3 days/acre (20 doses)	312.5	Soil application in two rings (1.63 kg/tree)	95.2	4.8 kg/3 days/acre (20 doses)
March	31 days	63	6.3 kg/3 days/acre (10 doses)	0		63.5	6.4 kg/3 days/acre (10 doses)
April-June	90 days	126	4.2 kg/3 days/acre (30 doses)	312.5	Soil application in two rings (1.63 kg/tree)	158.7	5.3 kg/3 days/acre (30 doses)

@Spray Borax 0.5% solution to check fruit drop.
Spray Planofix + 1% urea at flowering

4. Pomegranate (*Punica granatum*)

RD

Age	N	P	K	FYM
1st year	200 g/tree	100 g/tree	100 g/tree	25 kg/tree
2nd year	350	175	175	,,
3rd year	500	250	250	,,
4th year	800	400	400	100 kg/tree
5th year	Continue as above.			

Conventionally fertilizer is applied in two split doses during June – July and in October. Apply all the FYM in two splits, in early June and September.

RD Recommendation made for Tissue Culture (TC) high productivity Pomegranate

Fertilizer requirement for TC Pomegranate

Age of tree (months)	FYM kg/tree	N g/tree	P g/tree	K g/tree
18-Jan	20	375	187.5	166.2
19-24	10	62.5	62.5	146.3
25-36	30	625	250	250
4th year	40	625	250	250
5th year	50	625	250	250

RD TC planted at 2.1 m x 3 m High density Plantation. NPK kg/acre

Fertilizer requirement for TC Pomegranate				
Age of tree (months)	**FYM kg/acre**	**N kg/acre**	**P kg/acre**	**K kg/acre**
1-18 months	12840	245	123	109
19-24 months	6420	41	41	96
25-36 months (year 3)	19260	410	164	164
4th year	25680	410	164	164
5th year	32100	410	164	164

Ratio (1-18 months) 245:123:109 kg/acre NPK

1-18 months	Duration	N	P	K
1-90 days	90 days	0.1	0.1	0.1
91- 180	90	0.1	0.1	0.1
181-270	90	0.2	0.25	0.2
271-365	90	0.2	0.25	0.2
366-456	90	0.25	0.15	0.2
457-546	90	0.25	0.15	0.2

Fertigation schedule for the first 18 months (till first flowering)

1-8 Months	Duration	Urea kg/ acre	Schedule Urea	Phos-phoric Acid kg/acre	Schedule Phosphoric Acid	White MOP kg/acre	Schedule MOP
10-90 days	80 days	24.5	0.94 kg/3 days/acre (26 doses)	12.3	0.47 kg/3 days/acre (26 doses)	34.7	1.3 kg/3 days/acre (26 doses)
91- 180	90	24.5	0.82 kg/3 days/acre (30 doses)	12.3	0.41 kg/3 days/acre (30 doses)	34.7	1.2 kg/3 days/acre (30 doses)
181-270	90	49	1.6 kg/3 days/acre (30 doses)	30.8	1 kg/3 days/acre (30 doses)	52.2	1.7 kg/3 days/acre (30 doses)
271-365	90	49	1.6 kg/3 days/acre (30 doses)	30.8	1 kg/3 days/acre (30 doses)	52.2	1.7 kg/3 days/acre (30 doses)
366-456	90	61.3	2 kg/3 days/acre (30 doses)	18.5	0.62 kg/3 days/acre (30 doses)	52.2	1.7 kg/3 days/acre (30 doses)
457-546	90	61.3	2 kg/3 days/acre (30 doses)	18.5	0.62 kg/3 days/acre (30 doses)	52.2	1.7 kg/3 days/acre (30 doses)

Ratio (19-24 months) 41:41:96 kg/acre NPK

19-24 months (April-September)	Duration	N	P	K
547-591 days	45 days	0.34	0.5	0.2
592-636	45	0.33	0.5	0.2
637-681	45	0.33	0	0.3
682-726	45	0	0	0.3

Fertigation schedule for 19-24 months (yield formation and till harvest)

19-24 Months (April-September)	Duration	Urea kg/acre	Schedule Urea	Phosphoric Acid	Schedule Phosphoric Acid	White MOP kg/acre	Schedule MOP
546-591 days	45	30	2 kg/3 days/acre (15 doses)	39.4	2.6 kg/3 days/acre (15 doses)	32	2.1 kg/3 days/acre (15 doses)
592-636	45	30	2 kg/3 days/acre (15 doses)	39.4	2.6 kg/3 days/acre (15 doses)	32	2.1 kg/3 days/acre (15 doses)
637-681	45	30	2 kg/3 days/acre (15 doses)	0		48	3.2 kg/3 days/acre (15 doses)
682-726	45	0		0		48	3.2 kg/3 days/acre (15 doses)

Ratio (25-36 months; year 3) 410:164:164 kg/acre NPK.

25-36 month	Duration (days)	N	P	K
October -November 15	45	0.3	0.4	0.1
Nov 16-Dec 31	45	STRESS PERIOD		
Jan 1- Feb	58	0.3	0.3	0.2
March-May	90	0.3	0.3	0.3
June- July	90	0.1	0	0.3
Aug -Sept	60	0	0	0.1

Fertigation Schedule (25-36 months; Year 3 adapting for the *Bahar* treatment)

25-36 Months (April-September)	Duration (days)	Urea kg/ acre	Schedule Urea	Phos-phoric Acid kg/acre	Schedule Phosphoric Acid (52%P) g/wk/tree	White MOP kg/acre	Schedule MOP
October-November 15	45	267	12.1 kg/2 days/acre (22 doses)	126	5.7 kg/2 days/acre (22 doses)	27.4	1.1 kg/2 days/acre (22 doses)
Nov 16-Dec 31	45	**STRESS PERIOD for inducing Flower**					
Jan 1- Feb	58	267	9.2 kg/2 days/acre (29 doses)	94.5	3.3 kg/2 days/acre (29 doses)	54.8	1.9 kg/2 days/acre (29 doses)
March-May	90	267	5.9 kg/2 days/acre (45 doses)	94.5	2.1 kg/2 days/acre (45 doses)	82	1.8 kg/2 days/acre (45 doses)
June- July	90	89	1.98 kg/2 days/acre (45 doses)	0	0	82	1.8 kg/2 days/acre (45 doses)
Aug -Sept	60	0	0	0	0	27.4	0.91 kg/2 days/acre (30 doses)

5. Papaya (*Carica papaya*)

RD 110:100:200 NPK kg/acre

Ratio

Growth Stage	Duration	N	P	K
At planting	Basal Soil application	0	0.5	0
31-150 DAP	120 days	0.3	0.2	0.2
151-240 DAP	90 days	0.5	0.2	0.3
241-372 DAP	132 days	0.2	0.1	0.5

Fertilizers

- SSP
- Urea

- MAP
- MOP
- MgSO4
- Calcium nitrate

Fertigation Schedule

Growth Stage	Duration	Urea kg/acre	Schedule of Urea	SSP kg/acre	Schedule of SSP	MAP kg/acre	Schedule of MAP	MOP kg/acre	Schedule of MOP
At planting	Basal Soil application	0		312.5	Soil/pit application at planting	0		0	
31-150 DAP	120 days	67.9	1.7 kg/3 days/acre (40 doses)	0		32.8	0.82 kg/3 days/acre (40 doses)	66.8	1.67 kg/3 days/ acre (40 doses)
151-240 DAP	90 days	110.7	3.7 kg/3 days/acre (30 doses)	0		32.8	1.1 kg/3 days/acre (30 doses)	100.2	3.3 kg/3 days/ acre (30 doses)
241-372 DAP	132 days	43.4	1 kg/3 days/acre (44 doses)	0		16.4	0.4 kg/3 days/acre (44 doses)	167	3.8 kg/3 days/ acre (44 doses)

DAP = Days after planting the seedlings

MgSO4 and Calcium nitrate as follows:

MgSO4	30 kg/acre	100- 280 days	0.5 kg/3 days/acre (60 doses)
Ca(NO3)2	50 kg/acre	100-250 days	1 kg/3 days/acre (50 doses)

6. Citrus (*Citrus* spp.)

RD

Year	N	P	K kg/acre
year 1	22	18	11
year 2	31	18	13
year 3	33	34	54
year 4	42	38	65
year 5	51	50	76

Ratio (year 1 & 2)

Growth time	Duration	N	P	K
year 1 July-Sept	92 days	0.35	0.5	0.4
Jan-Apr	120 days	0.65	0.5	0.6
year 2 July-Sept	92 days	0.35	0.5	0.4
Jan-Apr	120 days	0.65	0.5	0.6

Fertigation schedule (Year 1 and 2)

Year	Growth Time	Duration	Urea kg/ acre	Schedule of Urea	Phosphoric Acid kg/acre	Schedule of Phosphoric Acid	White MOP kg/ acre	Schedule of White MOP
1	Jul-Sep	92 days	16.8	0.56 kg/3 days/acre (30 doses)	17.5	0.58 kg/3 days/acre (30 doses)	7.4	0.74 kg/3 days/acre (10 doses in Sept. only)
	Jan-Apr	120 days	31.2	0.78 kg/3 days/acre (40 doses)	17.5	0.44 kg/3 days/acre (40 doses)	11	0.55 kg/3 days/acre (20 doses in Mar-Apr only)
2	Jul-Sep	92 days	23.6	0.79 kg/3 days/acre (30 doses)	17.5	0.58 kg/3 days/acre (30 doses)	8.8	0.88 kg/3 days/acre (10 doses in Sept only)
	Jan-Apr	120 days	43.7	1.1 kg/3 days/acre (40 doses)	17.5	0.44 kg/3 days/acre (40 doses)	13.2	0.66 kg/3 days/acre (20 doses in Mar-Apr only)

1. Planting in June of the year

Year 3 (Fruiting Year)

RD 33:25:35 kg/acre NPK

Ratio

year 3	Growth Time	Duration	N	P	K
	1st irrigation after Stress	30 days	0.3	0.55	0.1
	Flower bud to Fruit set (31- 90 days)	60 days	0.3	0.2	0.2
	Early Fruit Development (91-120 days)	30 days	0.1	0	0
	Fruit Development (91-150 days)	60 days	0.2	0.25	0.4
	Fruit Maturity (151-210 days)	80 days	0.1		0.3

Fertigation schedule (year 3)

Year 3	Growth Time	Duration	Urea kg/acre	Schedule of urea	Ammo. Sulphate kg/acre	Schedule of Ammo. Sulphate	MAP kg/ acre	Schedule of MAP	Calcium nitrate kg/acre	Schedule of Cal nitrate	SOP kg/ acre	Schedule of SOP
	1-30 days (after stress period)	30 days	0		35.2	3.52 kg/3 days/acre (10 doses)	22.6	2.26 kg/3 days/acre (10 doses)	0		8	0.8 kg/3 days/acre (10 doses)
	31-90 days	60 days	20	1 kg/3 days/acre (20 doses)	0		8.8	0.44 kg/3 days/acre (20 doses)	0		15	0.75 kg/3 days/acre (20 doses)
	91-120 days	30 days	0		0		0		15	1.5 kg/3 days/acre (10 doses)	0	
	91-150 days	60 days	10	0.5 kg/3 days/acre (20 doses)	0		9.6	0.48 kg/3 days/acre (20 doses)	0		25	1.25 kg/3 days/acre (20 doses)
	151-210 days	80 days	10	0.4 kg/3 days/acre (25 doses)	0		0		0		22	0.88 kg/3 days/acre (25 doses)

1. $MgSO_4$ 20 kg/acre should be applied 0.5 kg/3 days/acre (31-150 days period ; 40 doses).
2. The schedule is applicable to either Mrig or Ambebahar option. (Fertigation begins with first irrigation after the stress period in both cases).
3. Micro nutrients are fertigated based on the requirement.
4. Study the fertilizer compatibility table while fertigation Ammonium sulphate, MAP and Calcium nitrate.

7. Guava (*Psidium guajava*)

A. Conventional Plantation (162 plants/acre at 5 x 5 m)

Year 1

RD 16:7:16NPK kg/acre (APAU recommendation)

Ratio

Growth Time	Duration	N	P	K
June-Aug	80 days	0.5	0.5	0.5
Sept-Oct	60 days	0.5	0.5	0.5

Fertigation Schedule

Year	Growth Time	Duration	Urea kg/ acre	Schedule of Urea	MAP kg/acre	Schedule of MAP	White MOP kg/ acre	Schedule of White MOP
1	June-Aug	80 days	16.3	0.63 kg/3 days/acre (26 doses)	5.3	0.20 kg/3 days/acre (26 doses)	13.5	0.52 kg/3 days/acre (26 doses)
	Sept-Oct	60 days	16.3	0.82 kg/3 days/acre (20 doses)	5.3	0.27 kg/3 days/acre (20 doses)	13.5	0.68 kg/3 days/acre (20 doses)

Flowers are clipped off in the first year

Year 2

RD 32:13:32 NPK kg/acre

Ratio

Growth Time	Duration	N	p	K
Dec-Jan	60 days	0.5	0.5	0.2
Feb-April 15	74 days			0.3
April end Pruning				
Pruning-July	90 days	0.5	0.5	0.2
Aug-Sept	45 days			0.3
Oct-Nov pruning				

Fertigation schedule

Year	Growth Time	Duration	Urea kg/ acre	Schedule of Urea	MAP kg/acre	Schedule of MAP	White MOP kg/ acre	Schedule of White MOP
2	Dec-Jan	60 days	32.4	1.6 kg/3 days/acre (20 doses)	10.7	0.54 kg/3 days/acre (20 doses)	10.7	0.54 kg/3 days/acre (20 doses)
	Feb-April 15	74 days					16	0.67 kg/3 days/acre (24 doses)
	April end	Pruning						
	Pruning-July	90 days	32.4	1.1 kg/3 days/acre (30 doses)	10.7	0.36 kg/3 days/acre (30 doses)	10.7	0.36 kg/3 days/acre (30 doses)
	Aug-Sept	45 days					16	1.1 kg/3 days/acre (15 doses)
	Oct-Nov	pruning						

1. Ca and Mg is fertigated as Calcium nitrate and MgSO4 both 1 kg each 3 times in February (every 10 days)in year 2.

2. Micro nutrient mixture–Multiplex 4 sprays at 2 ml/l concentration during regrowth after pruning and during fruit setting.

B. High density plantation (674plants/acre at 3 x 2 m)

Fruiting stage

RD 242:123:130 NPK kg/acre (Recommendation from Lucknow Institute)

Ratio

Growth Time	Duration	N	P	K
Immediately After Pruning (1st Irrigation to 60 days (June -Aug)	60 days	0.65	0.35	0.2
Flowering & Fruit Setting & Fruit Dev 60 to 120 days	60 days	0.35	0.5	0.45
Fruit Dev & Maturity 120 to 160 days	40 days	0.05	0.15	0.35

Fertigation Schedule

Growth Time	Duration	Urea kg/ acre	Schedule of Urea	MAP kg/acre	Schedule of MAP	White MOP kg/ acre	Schedule of MOP (white)
Immediately after Pruning (1st Irrigation to 60 days (June-July)	60 days	322.9	16.1 kg/3 days/acre (20 doses)	70.6	3.5 kg/3 days/acre (20 doses)	43.4	2.2 kg/3 days/acre (20 doses)
Flowering & Fruit Setting & Fruit Dev. 61 -120 days	60 days	157.5	7.9 kg/3 days/acre (20 doses)	101	5.2 kg/3 days/acre (20 doses)	97.7	4.9 kg/3 days/acre (20 doses)
Fruit Dev. & Maturity 120 to 160 days	40 days	18.4	1.4 kg/3 days/acre (13 doses)	30.3	2.3 kg/3 days/acre (13 doses)	76	5.8 kg/3 days/acre (13 doses)

1. Fertigate 1.25 kg/3 days Calcium nitrate every 3 days during flowering and fruit setting period (20 doses)
2. Fertigate 1 kg/3 days Magnesium sulphate every 3 days during flowering and fruit setting period (20 doses)
3. Multiplex @ 2 g /lit, 3 to 4 sprays after Pruning & at Fruit Setting
4. After every pruning Spray with Mn @ 5 g & Zn EDTA@ 2g and Fe + Copper @ 2.5 g of each to increase new flush & Control Browning of leaves
5. 90 days after fruit setting apply 90 kg Neem cake/tree apply to the base.
6. During Fruit setting to Fruit development stage Spray 2 % Urea plus ZnSo4 @ 4 g and Boric Acid @ 2 g.

8. Grapes (*Vitis* spp.)

RD 106:142:106 NPK kg/acre (Nasik Grapes association recommendation)

Ratios

Growth stages	N	P	K
April pruning /Back pruning			
Shoot growth (1-40 days)	0.15	0.28	0.07
Fruit Bud differentiation (41-60 days)	0.09	0.27	0.09
Cane Maturity and fruit development (61-120 days)	0.06	0.04	0.14
October Pruning/Forward Pruning			
Shoot growth (1-40 days)	0.22	0.06	0.07
Bloom to Shatter (41-55 days)	0.06	0.1	0.06
Berry growth and development (56-70 days)	0.06	0.1	0.06
Berry growth and development (71-105 days)	0.18	0.04	0.18
Ripening to harvest (106-135 days)	0.08	0.04	0.23
Harvest to foundation pruning (Rest period)	0.1	0.06	0.1

Fertigation schedule

Growth stage	Duration	Fertilizer	Total quantity kg/acre	Schedule kg/day/ acre
April pruning /Back pruning				
Shoot growth (1-40 days)	40 days	19:19:19	42	1.05
		MAP	52.4	1.31
		Urea	4	0.1
Fruit Bud differentiation (41-60 days)	20 days	Urea	4.4	0.22
		MAP	63.6	3.18
		SOP	19.2	0.96
Cane Maturity and fruit development (61-120 days)	60 days	19:19:19	33.6	0.56
		SOP	16.4	0.27
October Pruning/Forward Pruning				
Shoot growth (1-40 days)	40 days	Urea	34.8	0.87
		19:19:19	42	1.05
Bloom to Shatter (41-55 days)	15 days	Urea	6.8	0.45
		MAP	24.4	1.63
		SOP	12	0.8
Berry growth and development (56-70 days)	15 days	Urea	6.8	0.45
		MAP	24.4	1.63
		SOP	12	0.8
Berry growth and development (71-105 days)	35 days	Urea	28	0.8
		19:19:19	33.6	0.96
		SOP	25.6	0.73
Ripening to harvest (106-135 days)	30 days	Urea	16.8	0.56
		MAP	10	0.33
		SOP	49.6	1.65
Harvest to foundation pruning (Rest period)	20 days	Urea	19.6	0.98
		MAP	12	0.6
		SOP	20.8	1.04

1. Monitoring petiole content of nutrients and adjusting the nutrient quantities are essential for Grape production.
2. ICAR micro nutrient formulation for Grapes to be applied during the season.
3. Chloride containing fertilizers should not be used for Grapes.

9. Litchi (*Litchi chinensis*)

RD 135:36:170 NPK kg/acre (fruiting tree)

Ratios

Growth stages	Duration	N	P	K
June last week -July (immediately after pruning)	38 days	0.1	0.1	0
Aug-Sept.	60 days	0.4	0.5	0
Oct-Dec	90 days	0.4	0.4	0.15
January (complete stress)	31 days	0	0	0
February	28 days	0.05	0	0.15
March	31 days	0.05	0	0.3
April-May 15	45 days	0	0	0.4

Fertigation schedule

Growth Time	Duration	Urea kg/ acre	Schedule of Urea	MAP kg/acre	Schedule of MAP	White MOP kg/ acre	Schedule of MOP (white)
June last week-July (immediately after pruning)	38 days	27.8	2.3 kg/3 days/acre (12 doses)	5.9	0.5 kg/3 days/acre (12 doses)	0	
Aug-Sept.	60 days	109.6	5.48 kg/3 days/acre (20 doses)	29.5	1.48 kg/3 days/acre (20 doses)	0	
Oct-Dec	90 days	111.1	3.7 kg/3 days/acre (30 doses)	23.6	0.79 kg/3 days/acre (30 doses)	42.6	1.42 kg/3 days/acre (30 doses)
January (complete stress) 31 days	0		0		0		
February	28 days	13.9	1.54 kg/3 days/acre (9 doses)	0		42.6	4.73 kg/3 days/acre (9 doses)
March	31 days	13.9	1.39 kg/3 days/acre (10 doses)	0		85.2	8.52 kg/3 days/acre (10 doses)
April-May 15	45 days	0		0		68	4.53 kg/3 days/acre (15 doses)

Micro nutrient spray in November (2 times); and in February, March (2 times each month)

10. Sapota (*Manilkara zapota*)

RD

	NPK kg/acre
year 1	4:04:04
year 2	8:08:08
year 3	21:21:21
year 4	42:42:42
year 5	82:82:82

Ratios

Year 1-2

Growth period	Duration	N	P	K
June-August	90 days	0.5	0.5	0.5
Oct- Dec 15	75 days	0.5	0.5	0.5

Year 3-6

Growth period	Duration	N	P	K
April-May	60 days	0.5	0.5	0.2
June-August 15	75 days	0.3	0.5	0.4
Sept 15-Oct	45 days	0.2	0	0.4

Fertigation Schedule (year 1)

Growth Time	Duration	Urea kg/ acre	Schedule of Urea	Phosphoric Acid kg/acre	Schedule Phosphoric of Acid	White MOP kg/	Schedule of MOP (white) acre
June-August	90 days	4.3	0.36 kg/ 7 days/acre (12 doses)	3.8	0.32 kg/ 7 days/acre (12 doses)	3.3	0.28 kg/ 7 days/acre (12 doses)
Oct-Dec 15	75 days	4.3	0.43 kg/ 7 days/acre (10 doses)	3.8	0.38 kg/ 7 days/acre (12 doses)	3.3	0.33 kg/ 7 days/acre (12 doses)

Fertigation schedule (Year 5)

Growth Time	Duration	Urea kg/ acre	Schedule of Urea	Phosphoric Acid kg/acre	Schedule Phosphoric of Acid	White MOP kg/	Schedule of MOP
April-May	60 days	89	4.5 kg/3 days/acre (20 doses)	78.7	2.6 kg/3 days/acre (20 doses)	27.4	1.4 kg/3 days/acre (20 doses)
June-August 15	75 days	53.4	2.1 kg/3 days/acre (25 doses)	78.7	3.1 kg/3 days/acre (25 doses)	54.8	2.2 kg/3 days/acre (25 doses)
Sept 15-Oct	45 days	35.6	2.4 kg/3 days/acre (15 doses)	0	0	54.8	3.7 kg/3 days/acre (15 doses)

1. Micronutirent mixture two sprays @10 g/l ; one during early June and two in September
2. GA at 500 ppm at early fruiting time

11. Ber (*Ziziphu smauritania*)

RD

Year 1 &2	6:1:1 NPK kg/acre
Year 3&4	12:1:3 NPK kg/acre
Year 5 onwards	32:3:4 NPK kg/acre

Ratio

	Growth period	Duration	N	P	K
Year 1&2	May -June	60 days	0.5	0.5	0.5
	Sept-Oct	60 days	0.5	0.5	0.5
Year 3&4	May -June	60 days	0.5	0.5	0.5
	Sept-Oct	60 days	0.5	0.5	0.5
Year 5 onwards	May -June	60 days	0.7	0.6	0
	Sept-Oct	60 days	0.3	0.4	1

Fertigation schedule

Year 1&2

Growth Time	Duration	Urea kg/	Schedule of Urea	MAP kg/acre	Schedule of MAP	MOP kg/	Schedule of MOP acre
May-June	60 days	6.3	0.32 kg/3 days/acre (20 doses)	0.8	0.1 kg/3 days/ acre (8 doses in June)	0.85	0.0.11 kg/3 days/acre (8 doses in June)
Sept-Oct	60 days	6.3	0.32 kg/3 days/acre (20 doses)	0.8	0.1 kg/3 days/ acre (8 doses in Sept.)	0.85	0.0.11 kg/3 days/acre (8 doses in Sept.)

Year 3&4

Growth Time	Duration	Urea kg/	Schedule of Urea	MAP kg/acre	Schedule of MAP	MOP kg/acre	Schedule of MOP
May-June	60 days	5.9	0.3 kg/3 days/acre (20 doses)	0.8	0.1 kg/3 days/ acre (8 doses in June)	2.5	0.0.31 kg/3 days/acre (8 doses in June)
Sept-Oct	60 days	5.9	0.3 kg/3 days/acre (20 doses)	0.8	0.1 kg/3 days/ acre (8 doses in Sept.)	2.5	0.0.31 kg/3 days/acre (8 doses in Sept.)

Year 5 onwards

Growth Time	Duration	Urea kg/	Schedule of Urea	MAP kg/acre	Schedule of MAP	MOP kg/acre	Schedule of MOP
May-June	60 days	47.6	2.4 kg/3 days/acre (20 doses)	3	0.15 kg/3 days/acre (20 doses)	0	0
Sept-Oct	60 days	20.5	1 kg/3 days/acre (20 doses)	1.9	0.1kg/3 days /acre (20 doses)	6.7	0.34 kg/3 days/ acre (20 doses in Sept.)

12. Apple (*Malusdo mestica*)

RD High Density Apple planted at 4 x 2 m geometry. (505 trees/acre)

Age	N kg/acre	P kg/acre	K kg/acre
1 year	15	15	15
2	30	30	30
3	45	45	45
4	61	61	61
5	76	76	76
6	91	91	91
7	106	106	106
8	106	106	106

Ratio

Year	Growth stage	Duration	N	P	K
Year 1-3	March-August	184 days	1	1	1
Year 4	April	30 days	0.3	0.6	0
	May-June	61 days	0.7	0.4	0.3
	July-August	62 days	0	0	0.7
Year 5	April	30 days	0.3	0.6	0
	May-June	61 days	0.7	0.4	0.3
	July-August	62 days	0	0	0.7
Year 6	April	30 days	0.3	0.6	0
	May-June	61 days	0.7	0.4	0.3
	July-August	62 days	0	0	0.7
Year 7 onwards	April	30 days	0.3	0.6	0
	May-June	61 days	0.7	0.4	0.3
	July-August	62 days	0	0	0.7

Fertigation schedule

Year	Growth stage	Duration	Urea (kg/acre)	Schedule of Urea	Phosphoric Acid (kg/acre)	Schedule of Phosphoric Acid	MOP (kg/acre)	Schedule of MOP
Year 1-3	March-August	184 days	32.6	0.54 kg/3 days/ acre (60 doses)	29	0.48 kg/3 days/ acre (60 doses)	25	0.42 kg/3 days/ acre (60 doses)
Year 4	April	30 days	39.7	4 kg/3 days/acre (10 doses)	70.3	7 kg/3 days/acre (10 doses)	0	0
	May-June	61 days	92.7	4.6 kg/3 days/acre (20 doses)	46.8	2.3 kg/3 days/acre (20 doses)	30.6	1.5 kg/3 days/ acre (20 doses)
	July-August	62 days	0	0	0	0	71.3	3.7 kg/3 days/ acre (20 doses)
Year 5	April	30 days	49.5	5 kg/3 days/acre (10 doses)	87.6	8.8 kg/3 days/ acre (10 doses)	0	0
	May-June	61 days	115	5.8 kg/3 days/ acre (20 doses)	58.4	2.9 kg/3 days/ acre (20 doses)	38	1.9 kg/3 days/ acre (20 doses)
	July-August	62 days	0	0	0	0	88.8	4.4 kg/3 days/ acre (20 doses)
Year 6	April	30 days	59	5.9 kg/3 days/ acre (10 doses)	105	10.5 kg/3 days/ acre (10 doses)	0	0
	May-June	61 days	138.2	6.9 kg/3 days/ acre (20 doses)	70	3.5 kg/3 days/ acre (20 doses)	45.6	2.3 kg/3 days/ acre (20 doses)
	July-August	62 days	0	0	0	0	106.4	5.3 kg/3 days/ acre (20 doses)
Year 7 onward	April	30 days	69	6.9 kg/3 days/ acre (10 doses)	122	12.2 kg/3 days/ acre (10 doses)	0	0
	May-June	61 days	161	8.1 kg/3 days/ acre (20 doses)	81	4.1 kg/3 days/ acre (20 doses)	53	2.7 kg/3 days/ acre (20 doses)
	July-August	62 days	0	0	0	0	124	6.2 kg/3days/ acre (20 doses)

April corresponds to bud break and leaf growth; May –June is flower initiation and early fruit growth; and July –August is fruit enlargement and fruit maturity.

Ca and Mg is fertigated separately from above and given during the late April and May-June period.

Micro nutrient mixture is also applied during April – June period.

References

International Potash Institute (IPI) 2005. Fertigation Proceedings: Selected Papers on the IPI-NATESC-CAU-CAAS International Symposium on Fertigation, (ed. P. Imas and M. R. Price), Beijing, China, 20-24 September 2005.

Soman. P. 2021. Fertigation A novel method of applying crop nutrients.

Appendix

Conversion Tables

Table 1. Areas and weights, yields and application rates

From	To	Multiply by	From	To	Multiply by
Acre	Hectare	0.405	Hectare	acre	2.471
Kilogram	Pounds	2.205	Pound	kilogram	0.453
Gram	Ounces	0.035	Ounce	gram	28.35
Short ton	Metric ton (MT)	0.907	Metric ton (MT)	Short ton	1.1
Gallon (U.S)	Liters	3.785	Liter	gallon	0.26
kg/ha	Pound/acre	0.892	kg/ha	pound/acre	1.12
MT/ha	Pound/acre	892	pound/acre	MT/ha	0.001

Table 2. Elements and their oxides

Conversion factor from elements to their oxides Conversion factors from oxides to their elements

From	To	Multiply by	From	To	Multiply
N	NO_3^-	4.43	NO_3^-	N	0.23
N	NH_4^+	1.28	NH_4^+	N	0.82
P	P_2O_5	2.29	P_2O_5	P	0.44
K	K_2O	1.2	K_2O	K	0.83
Ca	Ca	O 1.4	CaO	Ca	0.71
Mg	MgO	1.66	MgO	Mg	0.6
S	SO_3	2.5	SO_3	S	0.4
S	SO_4	3	SO_4	S	0.33

Table 3. Atomic and equivalent weights

Element	Atomic Weight	Valence	Equivalent weight
H	1	1	1
N	14	1	14
K	39.1	1	39.1
Ca	40	2	20
Mg	24.3	2	12.15

Table 4. Soil Nutrient Ratings of Indian Soil

S.No.	Parameters/ Nutrients		Rating		
			Acidic	**Neutral**	**Alkaline**
1.	pH				
			<6.5	6.5-7.5	>7.5
2.	Electrical Conductivity		Safe	Moderate	Unsafe
		(dS/m)	<1.0	1.0-2.0	>2.0
	OC and Nutrients	Low	Medium	High	
3.	Organic Carbon (OC)	(%)	<0.5	0.5-0.75	>0.75
4.	Available Nitrogen (N)	kg/ha	<280	280-560	>560
5.	Available Phosphorus (P2O5)		<23	23-56	>56
6.	Available Potassium (K2O)		<130	130-336	>336
7.	Available Calcium (Ca)	ppm	<500	500-4100	>4100
8.	Available Magnesium (Mg)		<100	100-500	>500
9.	Available Sulphur (S)		<10	10-20	>20
10.	Available Iron (Fe)		<5	5-10	>10
11.	Available Manganese (Mg)		<5	5-10	>10
12.	Available Zinc (Zn)		<0.5	0.5-1.0	>1.0
13.	Available Copper (Cu)		<0.2	0.2-0.4	>0.4

Table 5. Sufficiency range of nutrients concentration in index tissue (leaf or Petiole) of major crops

Nutrient	Mango	Banana	Grapes	Orange	Pomeg-ranate	Apple	Coconut	Potato	Onion	Tomato	Cucumber	Sugar-cane
N, %	1.0-1.5	2.5-3.0	1.32-2.21	2.20-3.5	0.91-1.66	1.90-2.6	1.8-2.1	3.30-4.5	5.0-6.0	4.0-6.0	4.50-6.0	2.0-2.60
P,%	0.08-0.25	0.18-0.40	0.38-0.75	0.13-0.5	0.12-0.18	0.14-0.4	0.11-0.12	0.23-0.5	0.35-0.5	0.25-0.75	0.34-1.25	0.18-0.30
K,%	0.40-0.90	2.3-4.0	1.14-2.20	1.20-3.0	0.61-1.59	1.50-2.0	1.2-1.4	3.10-4.5	4.0-5.5	2.9-5.0	3.90-5.0	1.10-1.80
Ca, %	2.0-5.0	0.70-1.4	0.74-1.14	1.10-4.0	0.77-2.00	1.20-1.60	0.35-0.50	0.70-1.2	1.0-2.0	1.0-3.0	1.40-3.5	0.20-0.50
Mg, %	0.20-0.50	0.25-0.40	0.63-1.10	0.30-0.50	0.16-0.42	0.25-0.4	0.25-0.35	0.35-1.0	0.25-0.40	0.4-0.6	0.31-1.0	0.10-0.35
S, %	-	0.26-0.5	0.14-0.27		0.16-0.26	0.20-0.40	0.15-0.20	-	0.5-1.0	0.4-1.2	0.40-0.7	-
Fe, ppm	50-250	100-300	54-80	60-150	71-214	50-300	40-115	40-100	60-300	40-200	50-300	40-250
Mn, ppm	50-250	200-2000	76-174	25-200	29-89	25-200	60-120	40-250	50-250	40-250	50-300	25-400
Zn, ppm	20-200	13-50	53-132	25-200	14-72	20-100	60.00	20-50	25-10	20-50	25-100	20-100
Cu, ppm	7-50	6-30	5-10	6-100	29-72	6-50	12-13	-	15-35	5-20	7-20	5-15
B, ppm	-	-		25-100		25-50	8.00	25-75	22-60	25-60	25-60	4-30
Mo, ppm	-	-						-	-	-	0.8-3.3	0.05-4.0

Table 6. Fertilizer calculation table for different concentration solutions of N, P and K .

S.No.	Soluble Fertilizers	Nutrient Content (%)		Quantity of Fertilizer Required for Different Nutrient Concentration								
				PPM					%			
				1	10	50	100	1000	1	2	5	10
		Nutrient	%	Quantity (mg/lit)					Quantity (g/lit)			
1.	Urea	N	46	2	22	109	217	2174	22	43	109	217
2.	Single Super Phosphate (SSP)	P	16	6	63	313	625	6250	63	125	313	625
3.	Orthophosphoric Acid (H3PO4)	P	73	1	14	68	137	1370	14	27	68	137
4.	Potassium Chloride (MOP)	K	60	2	17	83	167	1667	17	33	83	167
5.	Potassium Sulphate (SOP)	K	50	2	20	100	200	2000	20	40	100	200
6.	Mono- Ammonium Phosphate (MAP)	N	12	8	83	417	833	8333	83	167	417	833
		P	61	2	16	82	164	1639	16	33	82	164
7.	Di- Ammonium Phosphate (DAP)	N	18	6	56	278	556	5556	56	111	278	556
		P	46	2	22	109	217	2174	22	43	109	217
8.	Mono potassium Phosphate	P	52	2	19	96	192	1923	19	38	96	192
		K	34	3	29	147	294	2941	29	59	147	294
10.	Potassium Nitrate	N	13	8	77	385	769	7692	77	154	385	769
		K	45	2	22	111	222	2222	22	44	111	222
11.	Complex fertilizer 19:19:19	N	19	5	53	263	526	5263	53	105	263	526
		P	19	5	53	263	526	5263	53	105	263	526
		K	19	5	53	263	526	5263	53	105	263	526
12.	Complex fertilizer 12:32:16	N	12	8	83	417	833	8333	83	167	417	833
		P	32	3	31	156	313	3125	31	63	156	313
		K	16	6	63	313	625	6250	63	125	313	625
13.	Complex fertilizer 10:26:26	N	10	10	100	500	1000	10000	100	200	500	1000
		P	26	4	38	192	385	3846	38	77	192	385
		K	26	4	38	192	385	3846	38	77	192	385
14.	Complex fertilizer 15;15;15	N	15	7	67	333	667	6667	67	133	333	667
		P	15	7	67	333	667	6667	67	133	333	667
		K	15	7	67	333	667	6667	67	133	333	667

Table 7. Fertilizer calculation table for different concentration of solutions of Micronutrients .

S. No.	Soluble Fertilizers	Nutrient contents (%)							Quantity of Fertilizer Required for Different Nutrient Concentration						
		Mg	B	Fe	Mn	Zn	Cu	S	PPM				%		
									1	10	100	500	0.1	0.5	1
									Quantity (mg/lit)				Quantity (g/lit)		
1.	MgSO4	16	-	-	-	-	-	13	6	63	625	3	6	31	63
2.	Borax	-	11	-	-	-	-	-	9	91	909	5	9	45	91
3.	FeSO4	-	-	19	-	-	-	12	5	53	526	3	5	26	53
4.	MnSO4	-	-	-	31	-	-	15	3	32	323	2	3	16	32
5.	ZnSO4	-	-	-	-	21	-	13	5	48	476	2	5	24	48
6.	CuSO4	-	-	-	-	-	24	13	4	42	417	2	4	21	42